LINEAR AND NONLINEAR DIFFERENTIAL EQUATIONS

ELLIS HORWOOD SERIES IN MATHEMATICS AND ITS APPLICATIONS

Series Editor: Professor G. M. BELL, Chelsea College, University of London
(and within the same series)

Statistics and Operational Research

Editor: B. W. CONOLLY, Chelsea College, University of London

Baldock, G. R. & Bridgeman, T. **MATHEMATICAL THEORY OF WAVE MOTION**
Beaumont, G. P. **INTRODUCTORY APPLIED PROBABILITY**
Burghes, D. N. & Wood, A. D. **MATHEMATICAL MODELS IN SOCIAL MANAGEMENT AND LIFE SCIENCES**
Burghes, D. N. **MODERN INTRODUCTION TO CLASSICAL MECHANICS AND CONTROL**
Burghes, D. N. & Graham, A. **CONTROL AND OPTIMAL CONTROL**
Burghes, D. N., Huntley, I. & McDonald, J. **APPLYING MATHEMATICS**
Butkovskiy, A. G. **GREEN'S FUNCTIONS AND TRANSFER FUNCTIONS HANDBOOK**
Butkovskiy, A. G. **STRUCTURE OF DISTRIBUTED SYSTEMS**
Chorlton, F. **TEXTBOOK OF DYNAMICS**
Chorlton, F. **VECTOR AND TENSOR METHODS**
Conolly, B. **TECHNIQUES IN OPERATIONAL RESEARCH**
Vol. 1: QUEUEING SYSTEMS
Vol. 2: MODELS, SEARCH, RANDOMIZATION
Dunning-Davies, J. **MATHEMATICAL METHODS FOR MATHEMATICS, PHYSICAL SCIENCE AND ENGINEERING**
Eason, G., Coles, C. W., Gettinby, G. **MATHEMATICS FOR THE BIOSCIENCES**
Exton, H. **HANDBOOK OF HYPERGEOMETRIC INTEGRALS**
Exton, H. **MULTIPLE HYPERGEOMETRIC FUNCTIONS**
Faux, I. D. & Pratt, M. J. **COMPUTATIONAL GEOMETRY FOR DESIGN AND MANUFACTURE**
Goult, R. J. **APPLIED LINEAR ALGEBRA**
Graham, A. **KRONECKER PRODUCTS AND MATRIX CALCULUS: WITH APPLICATIONS**
Graham, A. **MATRIX THEORY AND APPLICATIONS FOR ENGINEERS AND MATHEMATICIANS**
Griffel, D. H. **APPLIED FUNCTIONAL ANALYSIS**
Hoskins, R. F. **GENERALISED FUNCTIONS**
Hunter, S. C. **MECHANICS OF CONTINUOUS MEDIA, 2nd (Revised) Edition**
Huntley, I. & Johnson, R. M. **LINEAR AND NONLINEAR DIFFERENTIAL EQUATIONS**
Jones, A. J. **GAME THEORY**
Kemp, K. W. **COMPUTATIONAL STATISTICS**
Kosinski, W. **FIELD SINGULARITIES AND WAVE ANALYSIS IN CONTINUUM MECHANICS**
Marichev, O. I. **INTEGRALS OF HIGHER TRANSCENDENTAL FUNCTIONS**
Meek, B. L. & Fairthorne, S. **USING COMPUTERS**
Muller-Pfeiffer, E. **SPECTRAL THEORY OF ORDINARY DIFFERENTIAL OPERATORS**
Nonweiler, T. R. F. **COMPUTATIONAL MATHEMATICS: An Introduction to Numerical Analysis**
Oliviera-Pinto, F. **SIMULATION CONCEPTS IN MATHEMATICAL MODELLING**
Oliviera-Pinto, F. & Conolly, B. W. **APPLICABLE MATHEMATICS OF NON-PHYSICAL PHENOMENA**
Scorer, R. S. **ENVIRONMENTAL AERODYNAMICS**
Smith, D. K. **NETWORK OPTIMISATION PRACTICE: A Computational Guide**
Stoodley, K. D. C., Lewis, T. & Stainton, C. L. S. **APPLIED STATISTICAL TECHNIQUES**
Sweet, M. V. **ALGEBRA, GEOMETRY AND TRIGONOMETRY FOR SCIENCE STUDENTS**
Temperley, H. N. V. & Trevena, D. H. **LIQUIDS AND THEIR PROPERTIES**
Temperley, H. N. V. **GRAPH THEORY AND APPLICATIONS**
Twizell, E. H. **COMPUTATIONAL METHODS FOR PARTIAL DIFFERENTIAL EQUATIONS IN BIOMEDICINE**
Whitehead, J. R. **THE DESIGN AND ANALYSIS OF SEQUENTIAL CLINICAL TRIALS**

LINEAR AND NONLINEAR DIFFERENTIAL EQUATIONS

IAN HUNTLEY, B.A., Ph.D.

and

R. M. JOHNSON, B.Sc., A.F.I.M.A.

Department of Mathematics and Computing
Paisley College of Technology

ELLIS HORWOOD LIMITED
Publishers · Chichester

Halsted Press: a division of
JOHN WILEY & SONS
New York · Brisbane · Chichester · Toronto

First published in 1983 by

ELLIS HORWOOD LIMITED

Market Cross House, Cooper Street, Chichester, West Sussex, PO19 1EB, England

The publisher's colophon is reproduced from James Gillison's drawing of the ancient Market Cross, Chichester.

Distributors:

Australia, New Zealand, South-east Asia:
Jacaranda-Wiley Ltd., Jacaranda Press,
JOHN WILEY & SONS INC.,
G.P.O. Box 859, Brisbane, Queensland 40001, Australia

Canada:
JOHN WILEY & SONS CANADA LIMITED
22 Worcester Road, Rexdale, Ontario, Canada.

Europe, Africa:
JOHN WILEY & SONS LIMITED
Baffins Lane, Chichester, West Sussex, England.

North and South America and the rest of the world:
Halsted Press: a division of
JOHN WILEY & SONS
605 Third Avenue, New York, N.Y. 10016, U.S.A.

British Library Cataloguing in Publication Data
Huntley, I. D.
Linear and nonlinear differential equations. –
(Ellis Horwood series in mathematics and its applications)
1. Differential equations, Linear
2. Differential equations, Nonlinear
I. Title II. Johnson, R. M.
515.3'54 QA372

Library of Congress Card No. 82-25482

ISBN 0-85312-441-8 (Ellis Horwood Ltd., Publishers – Library Edn.)
ISBN 0-85312-583-X (Ellis Horwood Ltd., Publishers – Student Edn.)
ISBN 0-470-27413-1 (Halsted Press – Library Edn.)
ISBN 0-470-27420-4 (Halsted Press – Student Edn.)

Typeset in Press Roman by Ellis Horwood Ltd.
Printed in Great Britain by Butler & Tanner, Frome, Somerset.

Table of Contents

Authors' Preface 9

Part I: Systems of Linear Differential Equations

Chapter 1 Vector language 13
1.1 Introduction. 13
1.2 Linear dependence and independence 14
1.3 Vector spaces 16
1.4 Application of vector space algebra to the solution of $\dot{\mathbf{x}} = A\mathbf{x}$ 19

Chapter 2 Method of solution of $\dot{\mathbf{x}} = A\mathbf{x}$. 22
2.1 Introduction. 22
2.2 Eigenvalues and eigenvectors 22
2.3 Repeated eigenvalues 32
2.4 The matrix exponential 35
2.5 Complex eigenvalues and eigenvectors 44

Chapter 3 Geometrical considerations. 50
3.1 Introduction. 50
3.2 Initial value problems. 50
3.3 Geometrical interpretation of the solution of $\dot{\mathbf{x}} = A\mathbf{x}$ 57
3.4 Stability and the nature of solution curves 63

Chapter 4 Extensions to higher order 68
4.1 Introduction. 68
4.2 Reduction of higher order systems to $\dot{\mathbf{x}} = A\mathbf{x}$ 68
4.3 Undamped second order systems 72
4.4 Inhomogeneous equations. 78

Part II: Nonlinear Differential Equations – Graphical Methods

Chapter 5 Features of nonlinear differential equations. 93
5.1 Introduction. 93
5.2 Some properties of nonlinear differential equations 94

Chapter 6 Graphical methods of solution 98
6.1 Introduction. 98
6.2 The method of isoclines 98
6.3 Liénard's method 100

Chapter 7 Phase-plane analysis 104
7.1 Introduction. 104
7.2 The phase-plane 104
7.3 Graphical method for *xt* graph from phase-plane solution 111

Chapter 8 Linearisation techniques 115
8.1 Introduction. 115
8.2 Singular points 115

Part III: Nonlinear Differential Equations – Asymptotic Methods

Chapter 9 The methods of Poincaré and Lindstedt 137
9.1 Introduction. 137
9.2 Poincaré's method 138
9.3 Lindstedt's method 142
9.4 Forced oscillations. 146

Chapter 10 The multiple timescales method. 150
10.1 Introduction. 150
10.2 The method 150

Chapter 11 The method of Krylov and Bogoliubov 157
11.1 Introduction. 157
11.2 The method 157
11.3 Stability of limit cycles. 163

Chapter 12 Harmonic linearisation. 166
12.1 Introduction. 166
12.2 The method 166

Part IV: Oscillations in Certain Practical Situations

Chapter 13 Self-excited oscillations 171
13.1 Introduction. 171
13.2 Instability caused by friction 171
13.3 Galloping of transmission lines. 172
13.4 Vortex shedding 175
13.5 The simple castor 175
13.6 The valve oscillator 178

Chapter 14 Large nonlinearities 181
14.1 Introduction 181
14.2 Kipp oscillator experiment 181
14.3 Relaxation oscillations 182

References and further reading 188

Index 189

Authors' Preface

This textbook is intended to fill the gap in the existing literature between books on linear differential equations and those on advanced nonlinear differential equations. It is written as a text for mathematics students in universities, polytechnics and colleges, and for users of mathematics, such as control engineers, who require to be led from the study of linear differential equations to that of nonlinear differential equations. It also provides a useful starting point for those who require to read more advanced texts in the nonlinear field.

The level of presentation is appropriate to first and second year mathematics degree students or to readers who are not specialists in mathematics, such as engineers, scientists and others who encounter nonlinear dynamical systems. Such readers will require a mathematical background approximately equivalent to GCE A-level Pure Mathematics.

The book concentrates mainly on qualitative methods of solution and does not discuss numerical methods at all. Equations similar to those considered in this book are frequently solved in industrial applications by the use of specially written computer programs or by utilising general purpose computer packages. Qualitative methods, however, are important if they are able to provide an insight into the problem and reveal features which allow excessive computer use to be avoided. In addition, such methods can provide a check on computer solutions in specific cases.

Part I (Chapters 1–4) deals with linear differential equations by considering the first order system $\dot{\mathbf{x}} = A\,\mathbf{x}$, all homogeneous equations and many inhomogeneous equations being reducible to this form. This particular treatment is chosen since it is appropriate to some of the techniques used for nonlinear equations in later chapters. The minimum of vector algebra required for dealing with the solution of $\dot{\mathbf{x}} = A\,\mathbf{x}$ is covered in Chapter 1.

Part II (Chapters 5–8) introduces nonlinear differential equations and considers a number of graphical methods of solution. These methods do not always provide a complete solution but reveal many important properties of the solution, such as maximum values of parameters and conditions for stability.

Part III (Chapters 9–12) deals with asymptotic methods of solution, applied mainly to oscillating systems. Both steady-state and transient solutions are considered. Some of the techniques considered are comparatively new and are only available elsewhere in the literature in a form suitable for trained mathematicians.

The final Part (Chapters 13–14) presents oscillatory systems in a number of real life situations where both linear and nonlinear equations apply. A brief discussion and analysis is given in each case.

The 'dot' notation, $\dot{x} \equiv \mathrm{d}x/\mathrm{d}t$, is used throughout the book; other notations and abbreviations are defined as they occur in the text.

The authors are indebted to the many students at Paisley College who, over a number of years, have been on the receiving end of much of the material contained in this textbook.

Particular thanks are expressed to Janice Wallace for the considerable task of typing and correcting the manuscript.

Finally we are grateful to the Series Editor Professor G. Bell and to the staff of Ellis Horwood Ltd for their valuable assistance and encouragement.

I.D.H.
R.M.J.

Part I

Systems of linear differential equations

1

Vector language

1.1 INTRODUCTION

The analysis of systems of linear differential equations is simplified by the use of matrix and vector methods. For example, the system

$$\begin{aligned} \dot{x}_1 &= 5x_1 + 2x_2 + 2x_3 \\ \dot{x}_2 &= 2x_1 + 2x_2 + x_3 \\ \dot{x}_3 &= 2x_1 + x_2 + 2x_3 \end{aligned}$$

can be written as a single vector equation

$$\dot{\mathbf{x}} = A\,\mathbf{x}\ ,$$

where $\mathbf{x}$ is a *vector*, $\begin{bmatrix} x_1 \\ x_2 \\ x_3 \end{bmatrix}$, and

A is the *matrix*, $\begin{bmatrix} 5 & 2 & 2 \\ 2 & 2 & 1 \\ 2 & 1 & 2 \end{bmatrix}$.

The conversion to vector form is not done merely for notational convenience, although working with one equation instead of several is clearly an advantage. Important results relating to the solution of the differential equation can be obtained from common vector properties, and in many cases a geometrical interpretation of the solution is readily seen. In this chapter we shall discuss certain relevant vector concepts which are applicable to the solution of the system $\dot{\mathbf{x}} = A\,\mathbf{x}$.

1.2 LINEAR DEPENDENCE AND INDEPENDENCE

We define an n-DIMENSIONAL VECTOR as an ordered column of n elements,

i.e.
$$\mathbf{x} = \begin{bmatrix} x_1 \\ x_2 \\ x_3 \\ \vdots \\ x_n \end{bmatrix} .$$

A set of m vectors $\mathbf{v}_1, \mathbf{v}_2, \mathbf{v}_3, \ldots, \mathbf{v}_m$ is said to be LINEARLY DEPENDENT if one of these vectors can be expressed as a linear combination of the others. More precisely the set is linearly dependent if we can find constants $c_1, c_2, c_3, \ldots, c_m$, *not all zero,* such that

$$c_1\mathbf{v}_1 + c_2\mathbf{v}_2 + c_3\mathbf{v}_3 + \ldots + c_m\mathbf{v}_m = \mathbf{0} .$$

The set $\mathbf{v}_1$, $\mathbf{v}_2$, $\mathbf{v}_3$, . . . , $\mathbf{v}_m$ is said to be LINEARLY INDEPENDENT if the above equation can *only* be satisfied by taking all the constants $c_1, c_2, c_3, \ldots, c_m$ equal to zero.

Example 1.1

Show that the 3-dimensional vectors

$$\begin{bmatrix} 1 \\ 1 \\ 2 \end{bmatrix}, \quad \begin{bmatrix} 1 \\ 0 \\ 1 \end{bmatrix}, \text{ and } \begin{bmatrix} 0 \\ 0 \\ 1 \end{bmatrix} \text{ are linearly independent.}$$

We try to find c_1, c_2, c_3 such that

$$c_1 \begin{bmatrix} 1 \\ 1 \\ 2 \end{bmatrix} + c_2 \begin{bmatrix} 1 \\ 0 \\ 1 \end{bmatrix} + c_3 \begin{bmatrix} 0 \\ 0 \\ 1 \end{bmatrix} = \begin{bmatrix} 0 \\ 0 \\ 0 \end{bmatrix} ,$$

i.e.
$$\begin{aligned} c_1 + c_2 \quad\quad &= 0 \\ c_1 \quad\quad\quad &= 0 . \\ 2c_1 + c_2 + c_3 &= 0 \end{aligned}$$

Since $c_1 = 0$ then the first equation requires $c_2 = 0$ and substitution into the third equation implies $c_3 = 0$. Therefore the vector equation can only be satisfied by

$$c_1 = c_2 = c_3 = 0 .$$

Therefore the vectors are linearly independent.

Example 1.2

Are the three vectors

$$\begin{bmatrix} 1 \\ 1 \\ 2 \end{bmatrix}, \begin{bmatrix} 3 \\ 7 \\ 9 \end{bmatrix}, \text{ and } \begin{bmatrix} 5 \\ 9 \\ 13 \end{bmatrix}$$

linearly dependent or independent?

We try to find c_1, c_2 and c_3 such that

$$c_1 \begin{bmatrix} 1 \\ 1 \\ 2 \end{bmatrix} + c_2 \begin{bmatrix} 3 \\ 7 \\ 9 \end{bmatrix} + c_3 \begin{bmatrix} 5 \\ 9 \\ 13 \end{bmatrix} = \begin{bmatrix} 0 \\ 0 \\ 0 \end{bmatrix},$$

i.e.

$$c_1 + 3c_2 + 5c_3 = 0 \quad \text{(i)}$$

$$c_1 + 7c_2 + 9c_3 = 0 \quad . \quad \text{(ii)}$$

$$2c_1 + 9c_2 + 13c_3 = 0 \quad \text{(iii)}$$

Subtracting (ii) – (i) and (iii) – 2 × (i) gives

$$4c_2 + 4c_3 = 0$$

$$3c_2 + 3c_3 = 0 .$$

Therefore we can find a non-trivial solution

$$c_2 = -c_3 = k, \text{ arbitrary}, c_1 = 2k .$$

For example, taking $k = 1$, we have

$$2 \begin{bmatrix} 1 \\ 1 \\ 2 \end{bmatrix} + \begin{bmatrix} 3 \\ 7 \\ 9 \end{bmatrix} + (-1) \begin{bmatrix} 5 \\ 9 \\ 13 \end{bmatrix} = \begin{bmatrix} 0 \\ 0 \\ 0 \end{bmatrix} .$$

Therefore the vectors are linearly dependent.

[An alternative method is to discover whether the determinant

$$\begin{vmatrix} 1 & 3 & 5 \\ 1 & 7 & 9 \\ 2 & 9 & 13 \end{vmatrix}$$

is zero. Here this is certainly the case and so a non-trivial solution for c_1, c_2, c_3 exists.]

Exercise
Show that any *four* three-dimensional vectors are linearly dependent.

It follows from the above exercise that, *given* a set of 3 linearly independent three-dimensional vectors then *any other* three-dimensional vector can be expressed as a combination of the independent vectors. In particular if we take

$$\mathbf{i} = \begin{bmatrix} 1 \\ 0 \\ 0 \end{bmatrix}, \mathbf{j} = \begin{bmatrix} 0 \\ 1 \\ 0 \end{bmatrix}, \mathbf{k} = \begin{bmatrix} 0 \\ 0 \\ 1 \end{bmatrix}$$

as the independent vectors,

then the vector $\begin{bmatrix} a \\ b \\ c \end{bmatrix} = a\mathbf{i} + b\mathbf{j} + c\mathbf{k}$.

i, **j** and **k** are called a BASIS for the set of *all* three-dimensional vectors. (Any three independent vectors can be used as a basis.)

1.3 VECTOR SPACES

Let V be the set of all n-dimensional vectors $\mathbf{x}, \mathbf{y}, \mathbf{z}, \ldots$. Assuming that vector addition and scalar multiplication are defined in the usual way we have

$$\mathbf{x} + \mathbf{y} = \begin{bmatrix} x_1 \\ x_2 \\ \vdots \\ x_n \end{bmatrix} + \begin{bmatrix} y_1 \\ y_2 \\ \vdots \\ y_n \end{bmatrix} = \begin{bmatrix} x_1 + y_1 \\ x_2 + y_2 \\ \vdots \\ x_n + y_n \end{bmatrix} \text{ is a vector in } V$$

and
$$c\mathbf{x} = c\begin{bmatrix} x_1 \\ x_2 \\ \vdots \\ x_n \end{bmatrix} = \begin{bmatrix} cx_1 \\ cx_2 \\ \vdots \\ cx_n \end{bmatrix} \text{ is a vector in } V.$$

Also the following laws are satisfied for n-dimensional vectors.

(i) $\mathbf{x} + \mathbf{y} = \mathbf{y} + \mathbf{x}$,

(ii) $(\mathbf{x} + \mathbf{y}) + \mathbf{z} = \mathbf{x} + (\mathbf{y} + \mathbf{z})$,

(iii) there exists a member of V, $\mathbf{0} = \begin{bmatrix} 0 \\ 0 \\ \vdots \\ 0 \end{bmatrix}$,
such that $\mathbf{x} + \mathbf{0} = \mathbf{x}$,

(iv) there exists a member of V, $-\mathbf{x} = \begin{bmatrix} -x_1 \\ -x_2 \\ \vdots \\ -x_n \end{bmatrix}$,
such that $\mathbf{x} + (-\mathbf{x}) = \mathbf{0}$

(v) $1\,\mathbf{x} = \mathbf{x}$,

(vi) $\alpha(\beta\mathbf{x}) = (\alpha\beta)\mathbf{x}$,

(vii) $\alpha(\mathbf{x} + \mathbf{y}) = \alpha\mathbf{x} + \alpha\mathbf{y}$,

(viii) $(\alpha + \beta)\mathbf{x} = \alpha\mathbf{x} + \beta\mathbf{x}$.

We do not restrict our attention to vector sets since there are many non-vector situations where addition and scalar multiplication apply and where laws (i)–(viii) are satisfied.

Any set S where

(a) there exists a process (usually addition) of combining two members of S to form another member of S,

(b) there exists a process (usually scalar multiplication) of combining a number and a member of S to form another member of S,

(c) laws (i)–(viii) are satisfied,

is called a VECTOR SPACE. All the properties of a vector space can be deduced from the above definition.

Example 1.3

Show that the set S of all solutions of the differential equation

$$\frac{d^2y}{dt^2} + y = 0 ,$$

together with the processes of addition and scalar multiplication, is a vector space.

If $y = y_1$ and $y = y_2$ are solutions of $\frac{d^2y}{dt^2} + y = 0$, then we have to show that $y = y_1 + y_2$ and $y = cy_1$ are also solutions.

(a) $$\frac{d^2}{dt^2}(y_1 + y_2) + (y_1 + y_2) = \left\{\frac{d^2y_1}{dt^2} + y_1\right\} + \left\{\frac{d^2y_2}{dt^2} + y_2\right\}$$

$= 0$ because $y = y_1$ and $y = y_2$ are solutions.

Therefore $y = y_1 + y_2$ is a solution.

(b) $$\frac{d^2}{dt^2}(cy_1) + (cy_1) = c\left\{\frac{d^2y_1}{dt^2} + y_1\right\} = 0 .$$

Therefore $y = cy_1$ is a solution.

Now we have to show that laws (i)–(viii) are satisfied.

(i), (ii) and (v)–(viii) are clearly satisfied.

(iii) $y = 0$ is a solution of the differential equation and is therefore a member of S such that $y_1 + 0 = y_1$.

(iv) $y = -y_1$ is a solution since

$$\frac{d^2}{dt^2}(-y_1) + (-y_1) = -\left\{\frac{d^2y_1}{dt^2} + y_1\right\} = 0$$

Therefore $y = -y_1$ is in S such that $y_1 + (-y_1) = 0$.
Therefore S is a vector space.

Exercise

Show that the set S of all solutions of the system of differential equations $\dot{\mathbf{x}} = A\mathbf{x}$, together with the processes of addition and scalar multiplication, is a vector space.

Further definitions

The definition of linear dependence and linear independence given in Section 1.2

for vectors applies also to the elements of any vector space. For example, $y = \sin t$ and $y = \cos t$ are solutions of

$$\frac{d^2y}{dt^2} + y = 0 \ ,$$

(see Example 1.3).

They are independent solutions since $c_1 \sin t + c_2 \cos t = 0$ can only be satisfied *for all* t by taking $c_1 = c_2 = 0$. However, $y = \sin t$, $y = \cos t$ and $y = 2 \sin t - \cos t$ are linearly dependent solutions since

$$(-2)(\sin t) + (\cos t) + (2 \sin t - \cos t) = 0$$

for all t.

If a set $e_1, e_2, e_3, \ldots, e_n$ of linearly independent elements of a vector space S is such that *all* elements of S can be expressed in the form $c_1 e_1 + c_2 e_2 + \ldots + c_n e_n$ then the set $e_1, e_2, \ldots, e_n$ is said to be a BASIS for S.

Further we say that the vector space S has DIMENSION n. The DIMENSION of a vector space S is the least number of elements that we have to find in order to know *all* the elements of S. Alternatively the dimension of a vector space is the largest number of linearly independent elements that can be found.

For example, $e_1 = \sin t$, $e_2 = \cos t$ is a basis for the vector space of solutions of

$$\frac{d^2y}{dt^2} + y = 0$$

in Example 1.3. So all solutions of the differential equation can be written in the form $y = c_1 \sin t + c_2 \cos t$, the general solution. The dimension of this vector space is 2.

Note that the basis for a given vector space is not unique. *Any* set of n linearly independent elements of a vector space S of dimension n is a basis for S.

1.4 APPLICATION OF VECTOR SPACE ALGEBRA TO THE SOLUTION OF $\dot{\mathbf{x}} = A\mathbf{x}$

We have already noted that solutions of $\dot{\mathbf{x}} = A\mathbf{x}$ form a vector space. We state the following two theorems.

Theorem 1

If $\dot{\mathbf{x}} = A\mathbf{x}$ is a system of n differential equations in n unknown functions $x_1, x_2, \ldots, x_n$, then the dimension of the vector space of solutions is n.

Theorem 2 (Test for linear independence of solutions)
If $\mathbf{x} = \mathbf{f}_1(t)$, $\mathbf{x} = \mathbf{f}_2(t)$, . . . , $\mathbf{x} = \mathbf{f}_m(t)$ are solutions of $\dot{\mathbf{x}} = A\mathbf{x}$ and we can find a value of $t = t_0$ such that $\mathbf{f}_1(t_0)$, $\mathbf{f}_2(t_0)$, . . . , $\mathbf{f}_m(t_0)$ are linearly independent vectors, then $\mathbf{f}_1(t)$, $\mathbf{f}_2(t)$, . . . , $\mathbf{f}_m(t)$ are linearly independent solutions of $\dot{\mathbf{x}} = A\mathbf{x}$.

Example 1.4

$$\mathbf{x} = \begin{bmatrix} 0 \\ e^t \\ 0 \end{bmatrix}, \quad \mathbf{x} = \begin{bmatrix} e^{2t} \\ e^{2t} \\ 0 \end{bmatrix}, \quad \mathbf{x} = \begin{bmatrix} 0 \\ 0 \\ 1 \end{bmatrix}$$

are solutions of

$$\dot{\mathbf{x}} = \begin{bmatrix} 2 & 0 & 1 \\ 1 & 1 & 0 \\ 0 & 0 & 0 \end{bmatrix} \mathbf{x} \, .$$

Show that they form a basis for the vector space of all solutions and write down the general solution.

Let $\mathbf{f}_1(t) = \begin{bmatrix} 0 \\ e^t \\ 0 \end{bmatrix}$, $\mathbf{f}_2(t) = \begin{bmatrix} e^{2t} \\ e^{2t} \\ 0 \end{bmatrix}$, $\mathbf{f}_3(t) = \begin{bmatrix} 0 \\ 0 \\ 1 \end{bmatrix}$

and put $t = 0$.

Then $\mathbf{f}_1(0) = \begin{bmatrix} 0 \\ 1 \\ 0 \end{bmatrix}$, $\mathbf{f}_2(0) = \begin{bmatrix} 1 \\ 1 \\ 0 \end{bmatrix}$, $\mathbf{f}_3(0) = \begin{bmatrix} 0 \\ 0 \\ 1 \end{bmatrix}$

are clearly linearly independent vectors.

By Theorem 2, $\mathbf{x} = \mathbf{f}_1(t)$, $\mathbf{x} = \mathbf{f}_2\,(t)$, $\mathbf{x} = \mathbf{f}_3(t)$ are three linearly independent solutions. Theorem 1 tell us that the dimension of the vector space of solutions is 3. The note at the end of Section 1.3 shows that the three given solutions form a basis.

The general solution is

$$\mathbf{x} = c_1 \begin{bmatrix} 0 \\ e^t \\ 0 \end{bmatrix} + c_2 \begin{bmatrix} e^{2t} \\ e^{2t} \\ 0 \end{bmatrix} + c_3 \begin{bmatrix} 0 \\ 0 \\ 1 \end{bmatrix} .$$

Problems

1. Verify that $\mathbf{x} = \begin{bmatrix} 0 \\ e^t \end{bmatrix}$ and $\mathbf{x} = \begin{bmatrix} e^t \\ te^t \end{bmatrix}$ are solutions of

$$\dot{\mathbf{x}} = \begin{bmatrix} 1 & 0 \\ 1 & 1 \end{bmatrix} \mathbf{x} .$$

Show that they form a basis for the vector space of all solutions and write down the general solution.

2. By putting $x_1 = x,$

$$x_2 = \dot{x},$$

$$x_3 = \ddot{x},$$

show that the third order differential equation

$$\dddot{x} + \ddot{x} - \dot{x} - x = 0$$

can be written in the form

$$\begin{bmatrix} \dot{x}_1 \\ \dot{x}_2 \\ \dot{x}_3 \end{bmatrix} = \begin{bmatrix} 0 & 1 & 0 \\ 0 & 0 & 1 \\ 1 & 1 & -1 \end{bmatrix} \begin{bmatrix} x_1 \\ x_2 \\ x_3 \end{bmatrix} .$$

Use theorem 2 to show that $x = e^t, x = e^{-t}$ and $x = te^{-t}$ form a basis for the set of solutions of this third order differential equation.

3. Show that the set of solutions of the nonlinear differential equation $\ddot{x} + x^2 = 0$ does not form a vector space with respect to addition and scalar multiplication.

4. Show that the set of all polynomials of degree less than or equal to 3 forms a vector space of dimension 4.

2

Method of solution of $\dot{\mathbf{x}} = A\mathbf{x}$

2.1 INTRODUCTION

We have seen in Chapter 1 that the set of solutions of $\dot{\mathbf{x}} = A\mathbf{x}$, where A is an $n \times n$ matrix of constant elements, forms a vector space of dimension n. It is therefore necessary to obtain n linearly independent solutions, $\mathbf{x} = \mathbf{u}_1$, $\mathbf{x} = \mathbf{u}_2$, $\ldots$, $\mathbf{x} = \mathbf{u}_n$, which can be used as a basis for the set of solutions. The GENERAL SOLUTION of the system is then a linear combination of the basis vectors, that is

$$\mathbf{x} = c_1\mathbf{u}_1 + c_2\mathbf{u}_2 + \ldots + c_n\mathbf{u}_n$$

where $c_1, c_2, \ldots, c_n$ are arbitrary constants.

A number of methods are available for finding the general solution of $\dot{\mathbf{x}} = A\mathbf{x}$; the method developed in this chapter considers certain algebraic properties of the matrix A. In other words we transform the problem from calculus to matrix algebra and, once the algebraic properties are known, we can *write down* the general solution of the system of differential equations. This is analogous to the method used for solution of

$$\ddot{y} + a\dot{y} + by = 0$$

by consideration of the quadratic equation $m^2 + am + b = 0$, where the general solution can be written down after the quadratic equation has been solved.

2.2 EIGENVALUES AND EIGENVECTORS

The first order system $\dot{x} = ax$ is a special case of the system $\dot{\mathbf{x}} = A\mathbf{x}$ and has solution $x = c_1 e^{at}$ where c_1 is a constant. This suggests that we should try $\mathbf{x} = e^{\lambda t}\mathbf{v}$ as a solution for $\dot{\mathbf{x}} = A\mathbf{x}$, where λ is an unknown constant and $\mathbf{v}$ is an unknown vector whose elements are constants.

Substituting $\mathbf{x} = e^{\lambda t}\mathbf{v}$, $\dot{\mathbf{x}} = \lambda e^{\lambda t}\mathbf{v}$ into the system equation gives

$$\lambda e^{\lambda t}\mathbf{v} = A e^{\lambda t}\mathbf{v} ,$$

i.e.
$$A\mathbf{v} = \lambda\mathbf{v} . \qquad (2.1)$$

Our algebraic problem is to find λ and $\mathbf{v}$ to satisfy equation 2.1 so that we can write down *one* solution, $\mathbf{x} = e^{\lambda t}\,\mathbf{v}$, of the system $\dot{\mathbf{x}} = A\mathbf{x}$. We would hope that n independent solutions are obtainable in this way so that a general solution of the system may be found.

We require to consider in some detail the solution of the algebraic problem, equation 2.1. Firstly the following definitions are necessary.

Any vector $\mathbf{v} \neq 0$ satisfying equation 2.1 is called an EIGENVECTOR of the matrix A. The corresponding value of λ is called an EIGENVALUE of the matrix A.

To obtain eigenvalues and eigenvectors for any given square matrix A we rewrite equation 2.1 as

$$(A - \lambda I)\,\mathbf{v} = \mathbf{0} \tag{2.2}$$

where I is the unit matrix of appropriate size. Equation 2.2 represents n simultaneous equations in n unknowns, the elements of $\mathbf{v}$. We require $\mathbf{v} \neq \mathbf{0}$, and the condition for a non-trivial solution of equation 2.2 is that the determinant of the matrix $A - \lambda I$ should be zero,

i.e.
$$|A - \lambda I| = 0\ . \tag{2.3}$$

Equation 2.3 is a polynomial equation in λ of degree n and will have n solutions (including possible repetitions). To obtain the eigenvectors of A we solve equation 2.2 for each distinct value of λ.

Example 2.1

Find the eigenvalues and eigenvectors of the matrix

$$A = \begin{bmatrix} 1 & 1 \\ 4 & 1 \end{bmatrix} .$$

$$A - \lambda I = \begin{bmatrix} 1-\lambda & 1 \\ 4 & 1-\lambda \end{bmatrix} .$$

$$|A - \lambda I| = (1-\lambda)^2 - 4 = \lambda^2 - 2\lambda - 3 = 0$$

from equation 2.3. Therefore the eigenvalues are

$$\lambda_1 = -1,\ \lambda_2 = 3\ .$$

For $\lambda_1 = -1$ let $\mathbf{v}_1$ be the corresponding eigenvector, and equation 2.2 becomes

$$\begin{bmatrix} 2 & 1 \\ 4 & 2 \end{bmatrix} \mathbf{v}_1 = \mathbf{0}\ .$$

$\mathbf{v}_1$ may be obtained by writing $\mathbf{v}_1 = \begin{bmatrix} p_1 \\ q_1 \end{bmatrix}$

and finding a non-trivial solution of the two equations in p_1 and q_1. However, in this simple case we can see by inspection that

$$\mathbf{v}_1 = \begin{bmatrix} 1 \\ -2 \end{bmatrix}$$

is a solution.

Note that there is no *unique* eigenvector corresponding to $\lambda_1 = -1$ since any multiple of $\begin{bmatrix} 1 \\ -2 \end{bmatrix}$ is also an eigenvector. We can say that the set of eigenvectors corresponding to $\lambda_1 = -1$ forms a vector space of dimension 1. $\begin{bmatrix} 1 \\ -2 \end{bmatrix}$ is a basis for this vector space so that any member can be written $c_1 \begin{bmatrix} 1 \\ -2 \end{bmatrix}$, where c_1 is a constant.

For $\lambda_2 = 3$ equation 2.2 becomes

$$\begin{bmatrix} -2 & 1 \\ 4 & -2 \end{bmatrix} \mathbf{v}_2 = \mathbf{0} \ ,$$

and by inspection $\mathbf{v}_2 = \begin{bmatrix} 1 \\ 2 \end{bmatrix}$ is the basis for the set of eigenvectors corresponding to $\lambda_2 = 3$, i.e. $c_2 \begin{bmatrix} 1 \\ 2 \end{bmatrix}$ is an eigenvector for any constant c_2.

Example 2.2

Apply the results of Example 2.1 to solve the system of differential equations

$$\dot{x}_1 = x_1 + x_2 \ ,$$

$$\dot{x}_2 = 4x_1 + x_2 \ .$$

Writing the equation in the form $\dot{\mathbf{x}} = A\mathbf{x}$ we have

$$\begin{bmatrix} \dot{x}_1 \\ \dot{x}_2 \end{bmatrix} = \begin{bmatrix} 1 & 1 \\ 4 & 1 \end{bmatrix} \begin{bmatrix} x_1 \\ x_2 \end{bmatrix} ,$$

i.e. $$A = \begin{bmatrix} 1 & 1 \\ 4 & 1 \end{bmatrix}.$$

From Example 2.1 we have that the eigenvalues of A are $\lambda_1 = -1, \lambda_2 = 3$ with corresponding eigenvectors

$$\mathbf{v}_1 = \begin{bmatrix} 1 \\ -2 \end{bmatrix}, \quad \mathbf{v}_2 = \begin{bmatrix} 1 \\ 2 \end{bmatrix}.$$

Therefore $x = e^{-t} \begin{bmatrix} 1 \\ -2 \end{bmatrix}$ and $x = e^{3t} \begin{bmatrix} 1 \\ 2 \end{bmatrix}$

are solutions of $\dot{\mathbf{x}} = A\mathbf{x}$.

Theorem 2 in Section 1.4 can be used to show that these two solutions are independent, and the dimension of the vector space of solutions is 2 (Theorem 1). Thus the two solutions form a basis for the set of solutions and the general solution is given by

$$\mathbf{x} = c_1 e^{-t} \begin{bmatrix} 1 \\ -2 \end{bmatrix} + c_2 e^{3t} \begin{bmatrix} 1 \\ 2 \end{bmatrix}.$$

Example 2.3

Show that $\mathbf{v}_1 = \begin{bmatrix} 2 \\ -3 \\ 2 \end{bmatrix}, \ \mathbf{v}_2 = \begin{bmatrix} 0 \\ 1 \\ -1 \end{bmatrix}, \ \mathbf{v}_3 = \begin{bmatrix} 0 \\ 1 \\ 2 \end{bmatrix}$

are eigenvectors of the matrix

$$A = \begin{bmatrix} 1 & 0 & 0 \\ 2 & 3 & 1 \\ 0 & 2 & 4 \end{bmatrix}$$

and find the corresponding eigenvalues.

Since the eigenvectors are given we simply have to apply equation 2.1 for each vector.

$$A\mathbf{v}_1 = \begin{bmatrix} 2 \\ -3 \\ 2 \end{bmatrix} = 1\mathbf{v}_1 \ .$$

$$A\mathbf{v}_2 = \begin{bmatrix} 0 \\ 2 \\ -2 \end{bmatrix} = 2\mathbf{v}_2 \ .$$

$$A\mathbf{v}_3 = \begin{bmatrix} 0 \\ 5 \\ 10 \end{bmatrix} = 5\mathbf{v}_3 \ .$$

Therefore we have proved that the given vectors are eigenvectors of A corresponding to eigenvalues $\lambda_1 = 1, \lambda_2 = 2, \lambda_3 = 5$.

Two useful theorems are now proved.

Theorem 1
If λ_1 and λ_2 are eigenvalues of A such that $\lambda_1 \neq \lambda_2$ then the corresponding eigenvectors are linearly independent.

Proof
Let $\mathbf{v}_1$ and $\mathbf{v}_2$ be the corresponding eigenvectors. Therefore

$$A\mathbf{v}_1 = \lambda_1\mathbf{v}_1 \ , \tag{2.4}$$

$$A\mathbf{v}_2 = \lambda_2\mathbf{v}_2 \ . \tag{2.5}$$

If we make the proposition that $\mathbf{v}_1$ and $\mathbf{v}_2$ are *dependent*, i.e. $\mathbf{v}_2 = k\mathbf{v}_1$ for some constant $k \neq 0$, then equation 2.5 becomes

$$Ak\mathbf{v}_1 = \lambda_2 k\mathbf{v}_1$$

i.e.

$$A\mathbf{v}_1 = \lambda_2\mathbf{v}_1 \ . \tag{2.6}$$

Subtracting (2.6) from (2.4) gives

$$\mathbf{0} = (\lambda_1 - \lambda_2)\mathbf{v}_1 \ . \tag{2.7}$$

Now $\mathbf{v}_1 \neq \mathbf{0}$ (since it is an eigenvector), therefore equation 2.7 implies that $\lambda_1 = \lambda_2$. However, we are given that $\lambda_1 \neq \lambda_2$, and so the proposition $\mathbf{v}_2 = k\mathbf{v}_1$ must be false. Therefore $\mathbf{v}_1$ and $\mathbf{v}_2$ are linearly indpendent.

An important corollary of this theorem applies to the system $\dot{\mathbf{x}} = A\mathbf{x}$. If $\lambda_1 \neq \lambda_2$ then $\mathbf{x} = e^{\lambda_1 t}\mathbf{v}_1$ and $x = e^{\lambda_2 t}\mathbf{v}_2$ are independent solutions.

Theorem 2
If λ_1 is an eigenvalue of A and

$$A_1, A_2, A_3, \ldots, A_n \quad \text{(not all zero)}$$

are the cofactors of any row of the matrix $A - \lambda_1 I$, then

$$\mathbf{v}_1 = \begin{bmatrix} A_1 \\ A_2 \\ A_3 \\ \vdots \\ A_n \end{bmatrix} \quad \text{is an eigenvector of } A$$

corresponding to $\lambda = \lambda_1$.

Proof (Given, for convenience, for the 3 × 3 case)
Without loss of generality we assume that the 1st row of $A - \lambda_1 I$ has cofactors A_1, A_2, A_3, (not all zero).

$$\text{Writing } A - \lambda_1 I = \begin{bmatrix} a_1 & a_2 & a_3 \\ b_1 & b_2 & b_3 \\ c_1 & c_2 & c_3 \end{bmatrix}, \text{ we have}$$

$$a_1A_1 + a_2A_2 + a_3A_3 = |A - \lambda_1 I| = 0$$

by equation 2.3,

$$b_1A_1 + b_2A_2 + b_3A_3 = \begin{vmatrix} b_1 & b_2 & b_3 \\ b_1 & b_2 & b_3 \\ c_1 & c_2 & c_3 \end{vmatrix} = 0,$$

$$c_1A_1 + c_2A_2 + c_3A_3 = \begin{vmatrix} c_1 & c_2 & c_3 \\ b_1 & b_2 & b_3 \\ c_1 & c_2 & c_3 \end{vmatrix} = 0 .$$

Therefore $(A - \lambda_1 I) \begin{bmatrix} A_1 \\ A_2 \\ A_3 \end{bmatrix} = \mathbf{0}$ and $\begin{bmatrix} A_1 \\ A_2 \\ A_3 \end{bmatrix}$

is an eigenvector of A by equation 2.2.

Example 2.4

Without reference to Example 2.3 find the eigenvalues and eigenvectors of the matrix

$$A = \begin{bmatrix} 1 & 0 & 0 \\ 2 & 3 & 1 \\ 0 & 2 & 4 \end{bmatrix} .$$

Write down the general solution of the system $\dot{\mathbf{x}} = A\mathbf{x}$ where A is the above matrix.

$$|A - \lambda I| = \begin{vmatrix} 1-\lambda & 0 & 0 \\ 2 & 3-\lambda & 1 \\ 0 & 2 & 4-\lambda \end{vmatrix}$$

$$= (1-\lambda)(3-\lambda)(4-\lambda) - 2(1-\lambda)$$
$$= (1-\lambda)(\lambda-2)(\lambda-5)$$
$$= 0 \text{ for eigenvalues} .$$

Therefore $\lambda_1 = 1, \lambda_2 = 2, \lambda_3 = 5$.

For $\lambda_1 = 1$,

$$(A - \lambda_1 I)\mathbf{v}_1 = \begin{bmatrix} 0 & 0 & 0 \\ 2 & 2 & 1 \\ 0 & 2 & 3 \end{bmatrix} \mathbf{v}_1 = \mathbf{0}$$

for eigenvectors.

The cofactors of $A - \lambda_1 I$ are, for row 1,

$$A_1 = \begin{vmatrix} 2 & 1 \\ 2 & 3 \end{vmatrix} = 4,$$

$$A_2 = -\begin{vmatrix} 2 & 1 \\ 0 & 3 \end{vmatrix} = -6,$$

$$A_3 = \begin{vmatrix} 2 & 2 \\ 0 & 2 \end{vmatrix} = 4 .$$

Therefore $\begin{bmatrix} 4 \\ -6 \\ 4 \end{bmatrix}$ is an eigenvector of A.

Therefore we take $\mathbf{v}_1 = \begin{bmatrix} 2 \\ -3 \\ 2 \end{bmatrix}$ as the basis for the set of eigenvectors corresponding to $\lambda_1 = 1$.

For $\lambda_2 = 2$,

$$(A - \lambda_2 I)\mathbf{v}_2 = \begin{vmatrix} -1 & 0 & 0 \\ 2 & 1 & 1 \\ 0 & 2 & 2 \end{vmatrix} \mathbf{v}_2 = \mathbf{0}$$

for eigenvectors.

The first row cofactors are

$$A_1 = 0, \; A_2 = -4, \; A_3 = 4 .$$

Therefore take $\mathbf{v}_2 = \begin{bmatrix} 0 \\ 1 \\ -1 \end{bmatrix}$ corresponding to $\lambda_2 = 2$.

For $\lambda_3 = 5$,

$$(A - \lambda_3 I)\mathbf{v}_3 = \begin{bmatrix} -4 & 0 & 0 \\ 2 & -2 & 1 \\ 0 & 2 & -1 \end{bmatrix} \mathbf{v}_3 = \mathbf{0}$$

for eigenvectors.

The first row cofactors are

$$A_1 = 0, \; A_2 = 2, \; A_3 = 4 \; .$$

$$\text{Therefore take } \mathbf{v}_3 = \begin{bmatrix} 0 \\ 1 \\ 2 \end{bmatrix} \text{ corresponding to } \lambda_3 = 5 \; .$$

Since all the eigenvalues are distinct, the corollary of Theorem 1 implies that

$$\mathbf{x} = e^{t} \begin{bmatrix} 2 \\ -3 \\ 2 \end{bmatrix}, \; \mathbf{x} = e^{2t} \begin{bmatrix} 0 \\ 1 \\ -1 \end{bmatrix} \text{ and } \mathbf{x} = e^{5t} \begin{bmatrix} 0 \\ 1 \\ 2 \end{bmatrix}$$

are independent solutions of $\dot{\mathbf{x}} = A\mathbf{x}$. Thus the general solution of $\dot{\mathbf{x}} = A\mathbf{x}$ is

$$\mathbf{x} = c_1 e^{t} \begin{bmatrix} 2 \\ -3 \\ 2 \end{bmatrix} + c_2 e^{2t} \begin{bmatrix} 0 \\ 1 \\ -1 \end{bmatrix} + c_3 e^{5t} \begin{bmatrix} 0 \\ 1 \\ 2 \end{bmatrix}$$

Problems

1. Solve $\dot{x}_1 = 6x_1 + 8x_2$,

   $\dot{x}_2 = -3x_1 - 4x_2$.

2. Solve $\dot{\mathbf{x}} = \begin{bmatrix} 5 & 4 \\ -1 & 0 \end{bmatrix} \mathbf{x}$, where $\mathbf{x}$ is $\begin{bmatrix} x_1 \\ x_2 \end{bmatrix}$.

3. Solve $\dot{x}_1 = -x_2$,

   $\dot{x}_2 = 2x_1 - 3x_2$.

4. Solve $\dot{\mathbf{x}} = \begin{bmatrix} 0 & 0 & 1 \\ 3 & 7 & -9 \\ 0 & 2 & -1 \end{bmatrix} \mathbf{x}$, where $\mathbf{x}$ is $\begin{bmatrix} x_1 \\ x_2 \\ x_3 \end{bmatrix}$.

5. Show that $A = \begin{bmatrix} 6 & -2 & 2 \\ -2 & 5 & 0 \\ 2 & 0 & 7 \end{bmatrix}$ has eigenvalues $\lambda_1 = 3$, $\lambda_2 = 6$, $\lambda_3 = 9$ and hence solve $\dot{\mathbf{x}} = A\mathbf{x}$.

6. The TRACE of an $n \times n$ matrix A is defined as

$$\text{Trace (A)} = \sum_{i=1}^{n} a_{ii} ,$$

where a_{ii} is the element of A in the ith row and ith column (i.e. on the leading diagonal). Using equation 2.3, show that

$$\text{Trace (A)} = \sum_{i=1}^{n} \lambda_i ,$$

the sum of all eigenvalues of A.
[This can be used as a check for eigenvalue calculations.]

7. Given that $\lambda = \lambda_1$ and $\mathbf{v} = \mathbf{v}_1$ are an eigenvalue and corresponding eigenvector of a matrix A, show that λ_1^2 is an eigenvalue of A^2 with corresponding eigenvector $\mathbf{v}_1$.

Answers to problems

1. $$\mathbf{x} = c_1 \begin{bmatrix} 4 \\ -3 \end{bmatrix} + c_2 e^{2t} \begin{bmatrix} -2 \\ 1 \end{bmatrix}$$

2. $$\mathbf{x} = c_1 e^{t} \begin{bmatrix} 1 \\ -1 \end{bmatrix} + c_2 e^{4t} \begin{bmatrix} 4 \\ -1 \end{bmatrix}$$

3. $$\mathbf{x} = c_1 e^{-t} \begin{bmatrix} 1 \\ 1 \end{bmatrix} + c_2 e^{-2t} \begin{bmatrix} 1 \\ 2 \end{bmatrix}$$

4.

$$\mathbf{x} = c_1 e^t \begin{bmatrix} 1 \\ 1 \\ 1 \end{bmatrix} + c_2 e^{2t} \begin{bmatrix} 1 \\ 3 \\ 2 \end{bmatrix} + c_3 e^{3t} \begin{bmatrix} 1 \\ 6 \\ 3 \end{bmatrix}$$

5.

$$\mathbf{x} = c_1 e^{3t} \begin{bmatrix} 2 \\ 2 \\ -1 \end{bmatrix} + c_2 e^{6t} \begin{bmatrix} -1 \\ 2 \\ 2 \end{bmatrix} + c_3 e^{9t} \begin{bmatrix} 2 \\ -1 \\ 2 \end{bmatrix}$$

2.3 REPEATED EIGENVALUES

The method of solution of $\dot{\mathbf{x}} = A\mathbf{x}$ given in Section 2.2 applies to any nth order system *provided* that we can find n *independent* solutions. This is always possible when the matrix A has n *distinct* eigenvalues, Theorem 1, but problems may arise when there are repeated eigenvalues. It may then not be possible to find n independent eigenvectors.

Compare the following simple examples.

Example 2.5

Solve the system $\dot{\mathbf{x}} = \begin{bmatrix} 1 & 0 \\ 0 & 1 \end{bmatrix} \mathbf{x}$.

$|A - \lambda I| = (1 - \lambda)^2 = 0$ for eigenvalues, therefore $\lambda = 1$ (twice).

$$(A - \lambda I)\mathbf{v} = \begin{bmatrix} 0 & 0 \\ 0 & 0 \end{bmatrix} \mathbf{v} = \mathbf{0} \text{ for eigenvectors} .$$

Thus here we *can* choose two *independent* eigenvectors

$\mathbf{v}_1 = \begin{bmatrix} 1 \\ 0 \end{bmatrix}$, $\mathbf{v}_2 = \begin{bmatrix} 0 \\ 1 \end{bmatrix}$, and so the general solution is

$$\mathbf{x} = c_1 e^t \begin{bmatrix} 1 \\ 0 \end{bmatrix} + c_2 e^t \begin{bmatrix} 0 \\ 1 \end{bmatrix} ,$$

i.e. $x_1 = c_1 e^t$; $x_2 = c_2 e^t$.

[Note that this follows immediately from the original equations, $\dot{x}_1 = x_1$, $\dot{x}_2 = x_2$, which are uncoupled and easily solved.]

Example 2.6

Solve the system $\dot{\mathbf{x}} = \begin{bmatrix} 1 & 0 \\ 1 & 1 \end{bmatrix} \mathbf{x}$.

$|A - \lambda I| = (1-\lambda)^2 = 0$ for eigenvalues, therefore $\lambda = 1$ (twice).

$$(A - \lambda I)\mathbf{v} = \begin{bmatrix} 0 & 0 \\ 1 & 0 \end{bmatrix} \mathbf{v} = \mathbf{0} \text{ for eigenvectors} .$$

Here $\begin{bmatrix} 0 \\ 1 \end{bmatrix}$ is an eigenvector corresponding to $\lambda = 1$ but there are no other independent eigenvectors. Thus we are unable to form a basis for the set of solutions, i.e. we cannot write down the general solution of the system using eigenvalues alone. In Section 2.4 we shall see that

$$\mathbf{x} = e^t \begin{bmatrix} 0 \\ 1 \end{bmatrix} \quad \text{and} \quad \mathbf{x} = e^t \begin{bmatrix} 1 \\ t \end{bmatrix}$$

are independent solutions and thus the general solution is

$$\mathbf{x} = c_1 e^t \begin{bmatrix} 0 \\ 1 \end{bmatrix} + c_2 e^t \begin{bmatrix} 1 \\ t \end{bmatrix} .$$

The following theorem allows us to find the number of independent eigenvectors available for a given eigenvalue of a matrix A.

Theorem 3

If A is an $n \times n$ matrix and the matrix $A - \lambda_1 I$ has rank s (i.e. $A - \lambda_1 I$ has s linearly independent rows), then there are $n{-}s$ linearly independent eigenvectors corresponding to the eigenvalue $\lambda = \lambda_1$.

[$n{-}s$ is called the DEGENERACY of the matrix $A - \lambda_1 I$. Note that $n{-}s \geqslant 1$ since $|A - \lambda_1 I| = 0$.]

In Example 2.5, $\lambda_1 = 1$ (twice),

$$A - \lambda_1 I = \begin{bmatrix} 0 & 0 \\ 0 & 0 \end{bmatrix} .$$

Therefore $n = 2, s = 0, n - s = 2$ and there are two independent eigenvectors of A corresponding to $\lambda_1 = 1$.

In Example 2.6, $\lambda_1 = 1$ (twice),

$$A - \lambda_1 I = \begin{bmatrix} 0 & 0 \\ 1 & 0 \end{bmatrix} .$$

Therefore $n = 2, s = 1, n - s = 1$ and there is only one independent eigenvector corresponding to the repeated eigenvalue $\lambda_1 = 1$.

Problems

1. Prove Theorem 3. (Hint: consider the number of independent equations defined by $(A - \lambda_1 I)\mathbf{v} = \mathbf{0}$.)

2. If $\mathbf{v}_1$ and $\mathbf{v}_2$ are independent eigenvectors of a matrix A corresponding to a repeated eigenvalue λ_1, show that $a\mathbf{v}_1 + b\mathbf{v}_2$ is also an eigenvector of A corresponding to λ_1. If in addition it is given that the degeneracy of $A - \lambda_1 I$ is 2, deduce that $\mathbf{v}_1$ and $\mathbf{v}_2$ form a basis for the set of eigenvectors corresponding to eigenvalue λ_1.

3. Show that 7, 1 and 1 are the eigenvalues for both of the matrices

$$A_1 = \begin{bmatrix} 5 & 2 & 2 \\ 2 & 2 & 1 \\ 2 & 1 & 2 \end{bmatrix} \text{ and } A_2 = \begin{bmatrix} 7 & 2 & 2 \\ 0 & 1 & 1 \\ 0 & 0 & 1 \end{bmatrix} .$$

Find the basis for the set of eigenvectors corresponding to each value of λ for both A_1 and A_2.

Solve the system $\dot{\mathbf{x}} = A_1\mathbf{x}$.

Answers to problems

3.

For $A_1, \lambda_1 = 7$, basis is $\mathbf{v}_1 = \begin{bmatrix} 2 \\ 1 \\ 1 \end{bmatrix}$.

For $A_1, \lambda_2 = 1$, basis is $\mathbf{v}_1 = \begin{bmatrix} 0 \\ 1 \\ -1 \end{bmatrix}$, $\mathbf{v}_2 = \begin{bmatrix} 1 \\ -2 \\ 0 \end{bmatrix}$.

For $A_2, \lambda_1 = 7$, basis is $\mathbf{v}_1 = \begin{bmatrix} 1 \\ 0 \\ 0 \end{bmatrix}$.

For $A_2, \lambda_2 = 1$, basis is $\mathbf{v}_1 = \begin{bmatrix} 1 \\ -3 \\ 0 \end{bmatrix}$.

For $\dot{\mathbf{x}} = A_1\mathbf{x}$, $\mathbf{x} = c_1 e^{7t} \begin{bmatrix} 2 \\ 1 \\ 1 \end{bmatrix} + c_2 e^t \begin{bmatrix} 0 \\ 1 \\ -1 \end{bmatrix} + c_3 e^t \begin{bmatrix} 1 \\ -2 \\ 0 \end{bmatrix}$.

2.4 THE MATRIX EXPONENTIAL

We recall that the single equation $\dot{x} = ax$ has solution $x = c_1 e^{at}$ and that the function e^{at} can be expanded as an infinite series

$$e^{at} = 1 + at + \frac{a^2t^2}{2!} + \frac{a^3t^3}{3!} + \ldots .$$

This suggests that the system $\dot{\mathbf{x}} = A\mathbf{x}$ has solution $\mathbf{x} = e^{At}\mathbf{v}$ where the *matrix* e^{At} is defined by

$$e^{At} = I + At + \frac{A^2t^2}{2!} + \frac{A^3t^3}{3!} + \ldots . \tag{2.8}$$

We must prove that this solution does in fact satisfy $\dot{\mathbf{x}} = A\mathbf{x}$.

Differentiating equation 2.8 gives

$$\frac{d}{dt}(e^{At}) = A + A^2t + \frac{A^3t^2}{2!} + \frac{A^4t^3}{3!} + \ldots$$

$$= A\left(I + At + \frac{A^2t^2}{2!} + \frac{A^3t^3}{3!} + \ldots\right)$$

$$= Ae^{At} .$$

$$\text{Therefore} \quad \dot{\mathbf{x}} = \frac{\mathrm{d}}{\mathrm{d}t}(\mathrm{e}^{At}\mathbf{v}) = A\mathrm{e}^{At}\mathbf{v} = A\mathbf{x} \; .$$

It follows that $\mathbf{x} = \mathrm{e}^{At}\mathbf{v}$ is a solution of $\dot{\mathbf{x}} = A\mathbf{x}$ for any vector $\mathbf{v}$ with constant elements, and it is possible to write down the general solution in terms of infinite series. This rather inconvenient form of solution is simplified using the following theorems.

Theorem 4

$$\mathrm{e}^{\lambda It} = \mathrm{e}^{\lambda t}I \; .$$

Theorem 5
Provided that $AB = BA$ for $n \times n$ matrices A and B, then

$$\mathrm{e}^{At}\,\mathrm{e}^{Bt} = \mathrm{e}^{(A+B)t} \; .$$

(The proofs of Theorem 3 and Theorem 4 follow directly from the definition of the matrix exponential, e^{At}, and are left as an exercise for the reader.)

Theorem 6

$$\mathrm{e}^{At} = \mathrm{e}^{\lambda t}\,\mathrm{e}^{(A-\lambda I)t} \quad \text{for any value of } \lambda \; .$$

Proof

$$\mathrm{e}^{At} = \mathrm{e}^{\lambda It}\,\mathrm{e}^{(A-\lambda I)t} \quad \text{by Theorem 5, since } \lambda I(A - \lambda I) = (A - \lambda I)\lambda I \; .$$

$$\text{Therefore} \quad \mathrm{e}^{At} = \mathrm{e}^{\lambda t}\,\mathrm{e}^{(A-\lambda I)t} \quad \text{by Theorem 4.}$$

Theorem 6 allows us to write the solution of $\dot{\mathbf{x}} = A\mathbf{x}$ as

$$\mathbf{x} = \mathrm{e}^{\lambda t}\,\mathrm{e}^{(A-\lambda I)t}\mathbf{v}$$

i.e.

$$\mathbf{x} = \mathrm{e}^{\lambda t}\left[\mathbf{v} + (A - \lambda I)\mathbf{v}t + (A - \lambda I)^2\mathbf{v}\frac{t^2}{2!} + (A - \lambda I)^3\mathbf{v}\frac{t^3}{3!} + \ldots\right] \quad (2.9)$$

Further, equation 2.9 gives the solution of $\dot{\mathbf{x}} = A\mathbf{x}$ for *any* λ and $\mathbf{v}$. The usefulness of equation 2.9 is in the choice of λ and $\mathbf{v}$ such that the series becomes *finite.* Referring to equation 2.2 one obvious choice is to take $\lambda = \lambda_1$ to be an eigenvalue of A and $\mathbf{v} = \mathbf{v}_1$ to be a corresponding eigenvector. Then

$$(A - \lambda_1 I)\mathbf{v}_1 = \mathbf{0} \; ,$$

$$(A - \lambda_1 I)^2\mathbf{v}_1 = (A - \lambda_1 I)(A - \lambda_1 I)\mathbf{v}_1 = \mathbf{0} \; ,$$

$$(A - \lambda_1 I)^3\mathbf{v} = (A - \lambda_1 I)(A - \lambda_1 I)^2\mathbf{v} = \mathbf{0} \; , \qquad \text{etc.}$$

So equation 2.9 reduces to $\mathbf{x} = e^{\lambda_1 t}\mathbf{v}_1$, which *we already know* is a solution of $\dot{\mathbf{x}} = A\mathbf{x}$. Further, *if* we can find a vector $\mathbf{w}_1$ such that

$$(A - \lambda_1 I)\mathbf{w}_1 \neq \mathbf{0}$$

but $$(A - \lambda_1 I)^2\mathbf{w}_1 = (A - \lambda_1 I)^3\mathbf{w}_1 = \ldots = \mathbf{0}$$

then we have *another* solution corresponding to the eigenvalue λ_1,

which is $\mathbf{x} = e^{\lambda_1 t}(\mathbf{w}_1 + (A - \lambda_1 I)\mathbf{w}_1 t)$

[Note that $(A - \lambda_1 I)\mathbf{w}_1$ is *an eigenvector of A* corresponding to eigenvalue λ_1 since

$$(A - \lambda_1 I)\ \{(A - \lambda_1 I)\mathbf{w}_1\} = \mathbf{0}\ .$$

This provides a useful check.]

This technique is used to obtain two independent solutions corresponding to a repeated eigenvalue (*when* this is necessary).

The two solutions $e^{\lambda_1 t}\mathbf{v}_1$ and $e^{\lambda_1 t}(\mathbf{w}_1 + (A - \lambda_1 I)\mathbf{w}_1 t)$ are independent since, following Theorem 2, Chapter 1, we put $t = 0$ and form the two vectors $\mathbf{v}_1$ and $\mathbf{w}_1$ which are clearly independent ($\mathbf{v}_1$ is eigenvector, $\mathbf{w}_1$ is *not* eigenvector).

The method can be extended if necessary to get three independent solutions when an eigenvalue is repeated three times. Then we would try to find a vector $\mathbf{z}_1$ such that

$$(A - \lambda_1 I)^2\mathbf{z}_1 \neq 0,\ (A - \lambda_1 I)^3\mathbf{z}_1 = (A - \lambda_1 I)^4\mathbf{z}_1 = \ldots = \mathbf{0}\ ,$$

and so a third solution would be given by

$$\mathbf{x} = e^{\lambda_1 t}\left[\mathbf{z}_1 + (A - \lambda_1 I)\mathbf{z}_1 t + (A - \lambda_1 I)^2\mathbf{z}_1\frac{t^2}{2!}\right].$$

Example 2.7

We considered $\dot{\mathbf{x}} = \begin{bmatrix} 1 & 0 \\ 1 & 1 \end{bmatrix}\mathbf{x}$ in Example 2.6, and obtained

$$\lambda_1 = 1 \text{ (twice) and } \mathbf{v}_1 = \begin{bmatrix} 0 \\ 1 \end{bmatrix},$$

i.e. $$\mathbf{x} = e^t \begin{bmatrix} 0 \\ 1 \end{bmatrix} \text{ is one solution .}$$

To find a second solution, we try to find $\mathbf{w}_1$ such that

$$(A - \lambda_1 I)\mathbf{w}_1 \neq \mathbf{0}$$

but $$(A - \lambda_1 I)^2\mathbf{w}_1 = (A - \lambda_1 I)^3\mathbf{w}_1 = \ldots = \mathbf{0} \ ,$$

i.e. $$\begin{bmatrix} 0 & 0 \\ 1 & 0 \end{bmatrix} \mathbf{w}_1 \neq \mathbf{0} \ ,$$

but $$\begin{bmatrix} 0 & 0 \\ 1 & 0 \end{bmatrix}^2 \mathbf{w}_1 = \begin{bmatrix} 0 & 0 \\ 0 & 0 \end{bmatrix} \mathbf{w}_1 = \mathbf{0} \ .$$

A suitable choice is $\mathbf{w}_1 = \begin{bmatrix} 1 \\ 0 \end{bmatrix}$, and $\mathbf{w}_1 + (A - \lambda_1 I)\mathbf{w}_1 t$

is $$\begin{bmatrix} 1 \\ 0 \end{bmatrix} + \begin{bmatrix} 0 & 0 \\ 1 & 0 \end{bmatrix} \begin{bmatrix} 1 \\ 0 \end{bmatrix} t$$

$$= \begin{bmatrix} 1 \\ 0 \end{bmatrix} + \begin{bmatrix} 0 \\ 1 \end{bmatrix} t$$

$$= \begin{bmatrix} 1 \\ t \end{bmatrix} \ .$$

Thus the second solution is $\mathbf{x} = \mathrm{e}^t \begin{bmatrix} 1 \\ t \end{bmatrix}$.

[Note that $(A - \lambda_1 I)\mathbf{w}_1 = \begin{bmatrix} 0 \\ 1 \end{bmatrix}$ which *is* an eigenvector of A as expected.]

The general solution is

$$\mathbf{x} = c_1\mathrm{e}^t \begin{bmatrix} 0 \\ 1 \end{bmatrix} + c_2\mathrm{e}^t \begin{bmatrix} 1 \\ t \end{bmatrix} \ .$$

Example 2.8

$$\text{Solve } \dot{\mathbf{x}} = \begin{bmatrix} -3 & 1 & 0 \\ 0 & -3 & 1 \\ 4 & -8 & 2 \end{bmatrix} \mathbf{x} \, .$$

$$|A - \lambda I| = (\lambda+3)^2(2-\lambda) + 4 - 8(\lambda+3) = 0 \text{ for eigenvalues} \, .$$

Therefore $\lambda^3 + 4\lambda^2 + 5\lambda + 2 = 0$,

and so $(\lambda+1)^2(\lambda+2) = 0$.

Therefore $\lambda_1 = -1$ (twice) and $\lambda_3 = -2$.

For eigenvectors corresponding to the repeated eigenvalue $\lambda_1 = -1$ consider

$$(A - \lambda_1 I)\mathbf{v}_1 = \mathbf{0} \, .$$

Therefore $\begin{bmatrix} -2 & 1 & 0 \\ 0 & -2 & 1 \\ 4 & -8 & 3 \end{bmatrix} \mathbf{v}_1 = \mathbf{0}, \; \mathbf{v}_1 = \begin{bmatrix} 1 \\ 2 \\ 4 \end{bmatrix}$ (no other independent eigenvector) .

Therefore $\mathbf{x} = e^{-t} \begin{bmatrix} 1 \\ 2 \\ 4 \end{bmatrix}$ is a solution of $\dot{\mathbf{x}} = A\mathbf{x}$.

To obtain a second solution we must try to choose $\mathbf{w}_1$ such that

$$(A - \lambda_1 I)\mathbf{w}_1 \neq \mathbf{0}$$

but $(A - \lambda_1 I)^2\mathbf{w}_1 = (A - \lambda_1 I)^3\mathbf{w}_1 = \ldots = \mathbf{0}$,

i.e. $\begin{bmatrix} -2 & 1 & 0 \\ 0 & -2 & 1 \\ 4 & -8 & 3 \end{bmatrix} \mathbf{w}_1 \neq \mathbf{0}$,

but $\begin{bmatrix} -2 & 1 & 0 \\ 0 & -2 & 1 \\ 4 & -8 & 3 \end{bmatrix}^2 \mathbf{w}_1 = \begin{bmatrix} 4 & -4 & 1 \\ 4 & -4 & 1 \\ 4 & -4 & 1 \end{bmatrix} \mathbf{w}_1 = \mathbf{0}$.

A suitable choice for $\mathbf{w}_1$ is $\begin{bmatrix} 1 \\ 1 \\ 0 \end{bmatrix}$ and so

$$\mathbf{w}_1 + (A - \lambda_1 I)\mathbf{w}_1 t \quad \text{is}$$

$$\begin{bmatrix} 1 \\ 1 \\ 0 \end{bmatrix} + \begin{bmatrix} -2 & 1 & 0 \\ 0 & -2 & 1 \\ 4 & -8 & 3 \end{bmatrix} \begin{bmatrix} 1 \\ 1 \\ 0 \end{bmatrix} t = \begin{bmatrix} 1-t \\ 1-2t \\ -4t \end{bmatrix} .$$

Therefore $\mathbf{x} = e^{-t} \begin{bmatrix} 1-t \\ 1-2t \\ -4t \end{bmatrix}$ is another solution of $\dot{\mathbf{x}} = A\mathbf{x}$.

For the eigenvector corresponding to λ_3 consider

$$(A - \lambda_3 I)\mathbf{v}_3 = \mathbf{0} .$$

Therefore $\begin{bmatrix} -1 & 1 & 0 \\ 0 & -1 & 1 \\ 4 & -8 & 4 \end{bmatrix} \mathbf{v}_3 = 0, \ \mathbf{v}_3 = \begin{bmatrix} 1 \\ 1 \\ 1 \end{bmatrix} .$

Therefore $\mathbf{x} = e^{-2t} \begin{bmatrix} 1 \\ 1 \\ 1 \end{bmatrix}$ is the third solution of $\dot{\mathbf{x}} = A\mathbf{x}$.

The general solution is

$$\mathbf{x} = c_1 e^{-t} \begin{bmatrix} 1 \\ 2 \\ 4 \end{bmatrix} + c_2 e^{-t} \begin{bmatrix} 1-t \\ 1-2t \\ -4t \end{bmatrix} + c_3 e^{-2t} \begin{bmatrix} 1 \\ 1 \\ 1 \end{bmatrix} .$$

Example 2.9

Given the matrix $A = \begin{bmatrix} -2 & 2 & 2 \\ 2 & -5 & 1 \\ 2 & 1 & -5 \end{bmatrix}$,

(i) show that $\begin{bmatrix} 2 \\ 1 \\ 1 \end{bmatrix}$ is an eigenvector,

(ii) show that $a \begin{bmatrix} 0 \\ 1 \\ -1 \end{bmatrix} + b \begin{bmatrix} 1 \\ 0 \\ -2 \end{bmatrix}$ is an eigenvector for arbitrary values of the constants a and b.

State the corresponding eigenvalue in each case and write down the general solution of the system

$$\dot{\mathbf{x}} = A\mathbf{x} \ .$$

Using the definition $A\mathbf{v} = \lambda\mathbf{v}$ we have

$$\text{(i)} \ A\mathbf{v} = \begin{bmatrix} -2 & 2 & 2 \\ 2 & -5 & 1 \\ 2 & 1 & -5 \end{bmatrix} \begin{bmatrix} 2 \\ 1 \\ 1 \end{bmatrix} = \begin{bmatrix} 0 \\ 0 \\ 0 \end{bmatrix} = 0\,\mathbf{v} = \lambda\mathbf{v} \ .$$

Therefore $\begin{bmatrix} 2 \\ 1 \\ 1 \end{bmatrix}$ is an eigenvector for eigenvalue $\lambda = 0$.

$$\text{(ii)} \ A\mathbf{v} = \begin{bmatrix} -2 & 2 & 2 \\ 2 & -5 & 1 \\ 2 & 1 & -5 \end{bmatrix} \begin{bmatrix} b \\ a \\ -a-2b \end{bmatrix} = \begin{bmatrix} -6b \\ -6a \\ 12b+6a \end{bmatrix}$$

$$= -6 \begin{bmatrix} b \\ a \\ -a-2b \end{bmatrix} = \lambda\mathbf{v} \ .$$

Therefore $a \begin{bmatrix} 0 \\ 1 \\ -1 \end{bmatrix} + b \begin{bmatrix} 1 \\ 0 \\ -2 \end{bmatrix}$ is an eigenvector for eigenvalue $\lambda = -6$.

$\begin{bmatrix} 0 \\ 1 \\ -1 \end{bmatrix}$ and $\begin{bmatrix} 1 \\ 0 \\ -2 \end{bmatrix}$ are independent eigenvectors

corresponding to $\lambda = -6$ and form a basis for the set of such eigenvectors. $\lambda = -6$ is in fact a repeated eigenvalue.

Thus we *have* the required three independent solutions to form the basis for the set of solutions of $\dot{\mathbf{x}} = A\mathbf{x}$, and the general solution is

$$\mathbf{x} = c_1 \begin{bmatrix} 2 \\ 1 \\ 1 \end{bmatrix} + c_2 e^{-6t} \begin{bmatrix} 0 \\ 1 \\ -1 \end{bmatrix} + c_3 e^{-6t} \begin{bmatrix} 1 \\ 0 \\ -2 \end{bmatrix}.$$

Note that in this example there are two independent eigenvectors corresponding to $\lambda_2 = -6$, since

$$A - \lambda_2 I = \begin{bmatrix} 4 & 2 & 2 \\ 2 & 1 & 1 \\ 2 & 1 & 1 \end{bmatrix} \text{ has degeneracy 2 },$$

and it is *not necessary* to find another solution of the form $e^{\lambda_2 t}(\mathbf{w}_2 + (A - \lambda_2 I)\mathbf{w}_2 t)$. In fact it is *impossible* to find such a solution.

Problems

1. Solve $\dot{\mathbf{x}} = \begin{bmatrix} -4 & 1 \\ -1 & -2 \end{bmatrix} \mathbf{x}$.

2. Solve $\dot{\mathbf{x}} = \begin{bmatrix} 5 & -6 & -6 \\ -1 & 4 & 2 \\ 3 & -6 & -4 \end{bmatrix} \mathbf{x}$.

3. Solve $\dot{\mathbf{x}} = \begin{bmatrix} 1 & 3 & -2 \\ 0 & 7 & -4 \\ 0 & 9 & -5 \end{bmatrix} \mathbf{x}$.

4. Show that the eigenvalues of the matrix

$$A = \begin{bmatrix} -2 & -1 & k \\ 1 & -4 & 1 \\ 0 & 0 & -3 \end{bmatrix}$$

are identical and independent of the constant k. Show that the general solution of the system $\dot{\mathbf{x}} = A\mathbf{x}$ will contain terms $t^2 e^{-3t}$ except when $k = 1$.

Solve the system for the case $k = 1$.

Answers to problems

1. $$\mathbf{x} = c_1 e^{-3t} \begin{bmatrix} 1 \\ 1 \end{bmatrix} + c_2 e^{-3t} \begin{bmatrix} t \\ 1+t \end{bmatrix} .$$

2. $$\mathbf{x} = c_1 e^{2t} \begin{bmatrix} 2 \\ 1 \\ 0 \end{bmatrix} + c_2 e^{2t} \begin{bmatrix} 2 \\ 0 \\ 1 \end{bmatrix} + c_3 e^{t} \begin{bmatrix} 3 \\ -1 \\ 3 \end{bmatrix} .$$

3. $$\mathbf{x} = c_1 e^{t} \begin{bmatrix} 1 \\ 0 \\ 0 \end{bmatrix} + c_2 e^{t} \begin{bmatrix} 0 \\ 2 \\ 3 \end{bmatrix} + c_3 e^{t} \begin{bmatrix} -2t \\ -4t \\ 1-6t \end{bmatrix} .$$

4. $$\mathbf{x} = c_1 e^{-3t} \begin{bmatrix} 1 \\ 1 \\ 0 \end{bmatrix} + c_2 e^{-3t} \begin{bmatrix} 0 \\ 1 \\ 1 \end{bmatrix} + c e^{-3t} \begin{bmatrix} 1+t \\ t \\ 0 \end{bmatrix}$$

[Note that other forms of answer are possible depending on the choice of eigenvectors and the vectors **w**.]

2.5 COMPLEX EIGENVALUES AND EIGENVECTORS

Complex eigenvalues and eigenvectors can be handled using the techniques already known. Thus if A has eigenvalues $\alpha \pm j\beta$ with corresponding eigenvectors $\mathbf{s} \pm j\mathbf{t}$, then

$$\mathbf{x} = e^{(\alpha \pm j\beta)t}(\mathbf{s} \pm j\mathbf{t}) \text{ are solutions of } \dot{\mathbf{x}} = A\mathbf{x} .$$

The general solution would then include

$$c_1 e^{(\alpha+j\beta)t}(\mathbf{s}+j\mathbf{t}) + c_2 e^{(\alpha-j\beta)t}(\mathbf{s}-j\mathbf{t})$$

which can be simplified using $e^{j\beta t} = \cos \beta t + j \sin \beta t$ and written in terms of real functions. However, the real functions can be obtained directly, by using the following theorem.

Theorem 7
If $\mathbf{x}(t) = \mathbf{X}(t) + j\mathbf{Y}(t)$ is a solution of $\dot{\mathbf{x}} = A\mathbf{x}$ then $\mathbf{x} = \mathbf{X}(t)$ and $\mathbf{x} = \mathbf{Y}(t)$ are also solutions.

Proof
We are given that

$$\frac{\mathrm{d}}{\mathrm{d}t}(\mathbf{X}(t) + j\mathbf{Y}(t)) = A\left[\mathbf{X}(t) + j\mathbf{Y}(t)\right]$$

Therefore $\dot{\mathbf{X}} + j\dot{\mathbf{Y}} = A\mathbf{X} + jAY$

Equating real and imaginary parts gives $\dot{\mathbf{X}} = A\mathbf{X}$ and $\dot{\mathbf{Y}} = A\mathbf{Y}$ which proves the theorem.

Further for the complex solution $\mathbf{x} = e^{(\alpha+j\beta)t}(\mathbf{s}+j\mathbf{t})$ arising from eigenvalue $\alpha + j\beta$ and eigenvector $\mathbf{s} + j\mathbf{t}$ it is easily shown that the real and imaginary parts are *independent* solutions, so there is no need to consider the eigenvalue $\alpha - j\beta$ at all.

Example 2.10

$$\text{Solve } \dot{\mathbf{x}} = \begin{bmatrix} 0 & 1 \\ -1 & 0 \end{bmatrix} \mathbf{x} .$$

$$|A - \lambda I| = \lambda^2 + 1 = 0 \text{ for eigenvalues } \lambda = \pm j .$$

To get the eigenvectors we consider $(A - \lambda I)\mathbf{v} = \mathbf{0}$.

$$\text{For } \lambda_1 = j , \quad \begin{bmatrix} -j & 1 \\ -1 & -j \end{bmatrix} \mathbf{v}_1 = \mathbf{0} , \quad \mathbf{v}_1 = \begin{bmatrix} 1 \\ j \end{bmatrix} .$$

Therefore $\mathbf{x} = e^{jt} \begin{bmatrix} 1 \\ j \end{bmatrix}$ is a solution of $\dot{\mathbf{x}} = A\mathbf{x}$,

i.e. $\mathbf{x} = (\cos t + j \sin t) \begin{bmatrix} 1 \\ j \end{bmatrix}$ is a complex solution

and Theorem 7 tells us that the real and imaginary parts are independent solutions.

Therefore $\operatorname{Re}\left\{ e^{jt} \begin{bmatrix} 1 \\ j \end{bmatrix} \right\} = \begin{bmatrix} \cos t \\ -\sin t \end{bmatrix}$ and $\operatorname{Im}\left\{ e^{jt} \begin{bmatrix} 1 \\ j \end{bmatrix} \right\} = \begin{bmatrix} \sin t \\ \cos t \end{bmatrix}$

are independent solutions.

$$\mathbf{x} = c_1 \begin{bmatrix} \sin t \\ \cos t \end{bmatrix} + c_2 \begin{bmatrix} \cos t \\ -\sin t \end{bmatrix}$$ is the required general

solution.

Example 2.11

Solve $\dot{\mathbf{x}} = \begin{bmatrix} 1 & 0 & 0 \\ 0 & 0 & 2 \\ 0 & -2 & 0 \end{bmatrix} \mathbf{x}$.

$|A - \lambda I| = (1-\lambda)\lambda^2 + 4(1-\lambda) = (1-\lambda)(\lambda^2+4) = 0$ for eigenvalues .

Therefore $\lambda_1 = 1,\ \lambda_2 = 2j,\ \lambda_3 = -2j$.

For $\lambda_1 = 1$, $\begin{bmatrix} 0 & 0 & 0 \\ 0 & -1 & 2 \\ 0 & -2 & -1 \end{bmatrix} \mathbf{v}_1 = \mathbf{0},\ \mathbf{v}_1 = \begin{bmatrix} 1 \\ 0 \\ 0 \end{bmatrix}$.

Therefore $\mathbf{x} = e^t \begin{bmatrix} 1 \\ 0 \\ 0 \end{bmatrix}$ is one solution.

For $\lambda_2 = 2j$, $\begin{bmatrix} 1-2j & 0 & 0 \\ 0 & -2j & 2 \\ 0 & -2 & -2j \end{bmatrix} \mathbf{v}_2 = \mathbf{0}$, $\mathbf{v}_2 = \begin{bmatrix} 0 \\ 1 \\ j \end{bmatrix}$.

Therefore $\mathbf{x} = e^{2jt} \begin{bmatrix} 0 \\ 1 \\ j \end{bmatrix}$

$$= (\cos 2t + j \sin 2t) \begin{bmatrix} 0 \\ 1 \\ j \end{bmatrix}$$

$$= \begin{bmatrix} 0 \\ \cos 2t \\ -\sin 2t \end{bmatrix} + j \begin{bmatrix} 0 \\ \sin 2t \\ \cos 2t \end{bmatrix} \text{ is a complex solution .}$$

By Theorem 7,

$$\mathbf{x} = \begin{bmatrix} 0 \\ \cos 2t \\ -\sin 2t \end{bmatrix} \text{ and } \mathbf{x} = \begin{bmatrix} 0 \\ \sin 2t \\ \cos 2t \end{bmatrix} \text{ are solutions .}$$

Therefore general solution is

$$\mathbf{x} = c_1 e^t \begin{bmatrix} 1 \\ 0 \\ 0 \end{bmatrix} + c_2 \begin{bmatrix} 0 \\ \cos 2t \\ -\sin 2t \end{bmatrix} + c_3 \begin{bmatrix} 0 \\ \sin 2t \\ \cos 2t \end{bmatrix} .$$

Example 2.12

Solve $\dot{\mathbf{x}} = \begin{bmatrix} -3 & 4 \\ -2 & 1 \end{bmatrix} \mathbf{x}$.

$|A - \lambda I| = (\lambda + 3)(\lambda - 1) + 8 = \lambda^2 + 2\lambda + 5 = 0$ for eigenvalues.

Therefore $\lambda_1 = -1 + 2j, \ \lambda_2 = -1 - 2j$.

For $\lambda_1 = -1 + 2j$,

$$(A - \lambda_1 I)\mathbf{v}_1 = \begin{bmatrix} -2-2j & 4 \\ -2 & 2-2j \end{bmatrix} \mathbf{v}_1 = \mathbf{0}, \quad \mathbf{v}_1 = \begin{bmatrix} 2 \\ 1+j \end{bmatrix} .$$

Therefore $\mathbf{x} = \mathrm{e}^{(-1+2j)t} \begin{bmatrix} 2 \\ 1+j \end{bmatrix}$

$$= \mathrm{e}^{-t}(\cos 2t + j \sin 2t) \begin{bmatrix} 2 \\ 1+j \end{bmatrix} \text{ is one solution .}$$

By Theorem 7, $\mathbf{x} = \mathrm{e}^{-t} \begin{bmatrix} 2\cos 2t \\ \cos 2t - \sin 2t \end{bmatrix}$

and $\mathbf{x} = \mathrm{e}^{-t} \begin{bmatrix} 2\sin 2t \\ \cos 2t + \sin 2t \end{bmatrix}$ are solutions.

Therefore general solution is

$$\mathbf{x} = c_1 \mathrm{e}^{-t} \begin{bmatrix} 2\cos 2t \\ \cos 2t - \sin 2t \end{bmatrix} + c_2 \mathrm{e}^{-t} \begin{bmatrix} 2\sin 2t \\ \cos 2t + \sin 2t \end{bmatrix} .$$

Problems

1. Solve $\dot{\mathbf{x}} = \begin{bmatrix} -2 & -5 \\ 1 & 2 \end{bmatrix} \mathbf{x}$.

2. Solve $\dot{\mathbf{x}} = \begin{bmatrix} 0 & 1 & 1 \\ -1 & 0 & 1 \\ -1 & -1 & 0 \end{bmatrix} \mathbf{x}$.

3. Solve $\dot{\mathbf{x}} = \begin{bmatrix} -3 & 0 & 2 \\ 1 & -1 & 0 \\ -2 & -1 & 0 \end{bmatrix} \mathbf{x}$.

4.

Show that $\begin{bmatrix} 0 \\ 1 \\ -1+j \end{bmatrix}$ is an eigenvector of the matrix

$$A = \begin{bmatrix} 1 & 0 & 0 \\ 0 & 1 & 1 \\ 0 & -2 & -1 \end{bmatrix}.$$

Hence solve $\dot{\mathbf{x}} = A\mathbf{x}$.

Answers to problems

1.

$$\mathbf{x} = \begin{bmatrix} (c_2 - 2c_1)\cos t - (c_1 + 2c_2)\sin t \\ c_1 \cos t + c_2 \sin t \end{bmatrix}$$

2.

$$\mathbf{x} = c_1 \begin{bmatrix} 1 \\ -1 \\ 1 \end{bmatrix} + c_2 \begin{bmatrix} \cos\sqrt{3}t + \sqrt{3}\sin\sqrt{3}t \\ 2\cos\sqrt{3}t \\ \cos\sqrt{3}t - \sqrt{3}\sin\sqrt{3}t \end{bmatrix}$$

$$+ c_3 \begin{bmatrix} \sin\sqrt{3}t - \sqrt{3}\cos\sqrt{3}t \\ 2\sin\sqrt{3}t \\ \sin\sqrt{3}t + \sqrt{3}\cos\sqrt{3}t \end{bmatrix}$$

3.

$$\mathbf{x} = c_1 e^{-2t} \begin{bmatrix} 2 \\ -2 \\ 1 \end{bmatrix} + c_2 e^{-t} \begin{bmatrix} -\sqrt{2}\sin\sqrt{2}t \\ \cos\sqrt{2}t \\ -\cos\sqrt{2}t - \sqrt{2}\sin\sqrt{2}t \end{bmatrix}$$

$$+ c_3 e^{-t} \begin{bmatrix} \sqrt{2}\cos\sqrt{2}t \\ \sin\sqrt{2}t \\ \sqrt{2}\cos\sqrt{2}t - \sin\sqrt{2}t \end{bmatrix}$$

4.

$$\mathbf{x} = c_1 e^t \begin{bmatrix} 1 \\ 0 \\ 0 \end{bmatrix} + c_2 \begin{bmatrix} 0 \\ -\cos t - \sin t \\ 2\sin t \end{bmatrix} + c_3 \begin{bmatrix} 0 \\ \cos t - \sin t \\ -2\cos t \end{bmatrix}$$

3

Geometrical considerations

3.1 INTRODUCTION

This chapter is concerned mainly with two-dimensional systems of differential equations

$$\dot{x}_1 = ax_1 + bx_2$$
$$\dot{x}_2 = cx_1 + dx_2 \ ,$$

and the interpretation of their solutions as curves in the x_1x_2 plane. The nature of these curves depends on the eigenvalues of the matrix $A = \begin{bmatrix} a & b \\ c & d \end{bmatrix}$ and the properties derived will be applied when certain nonlinear systems are considered in Chapter 8.

As a preliminary to the geometry of the x_1x_2 plane we apply the methods studied in Chapter 2 to the solution of initial value problems.

3.2 INITIAL VALUE PROBLEMS

Since differential equations usually are descriptions of physical systems, it is likely that the values of the dependent variables are known before the solution begins, i.e. at some time $t = t_0$ (usually $t_0 = 0$). This is called an INITIAL CONDITION. If we add to the system of differential equations $\dot{\mathbf{x}} = A\mathbf{x}$ the initial condition that $\mathbf{x} = \mathbf{x}_0$ when $t = t_0$ we have an INITIAL VALUE PROBLEM (IVP). We firstly obtain the general solution to $\dot{\mathbf{x}} = A\mathbf{x}$ and then use the condition $\mathbf{x}(t_0) = \mathbf{x}_0$ to evaluate the constants $c_1, c_2, \ldots$.

For instance, we have previously solved $\dot{\mathbf{x}} = \begin{bmatrix} 5 & 4 \\ -1 & 0 \end{bmatrix} \mathbf{x}$ and know that the general solution is

$$\mathbf{x} = c_1 e^t \begin{bmatrix} 1 \\ -1 \end{bmatrix} + c_2 e^{4t} \begin{bmatrix} 4 \\ -1 \end{bmatrix},$$

(Problem 2, Section 2.2).

Now consider the added condition that $\mathbf{x}(0) = \begin{bmatrix} 1 \\ 0 \end{bmatrix}$.

Substituting $t = 0$ into the general solution gives

$$\mathbf{x}(0) = c_1 e^0 \begin{bmatrix} 1 \\ -1 \end{bmatrix} + c_2 e^0 \begin{bmatrix} 4 \\ -1 \end{bmatrix} = \begin{bmatrix} 1 \\ 0 \end{bmatrix},$$

i.e. $\qquad c_1 + 4c_2 = 1$

$\qquad -c_1 - c_2 = 0$.

Therefore $c_1 = -\frac{1}{3}, \ c_2 = \frac{1}{3}$.

Thus the solution to the initial value problem is

$$\mathbf{x} = -\tfrac{1}{3} e^t \begin{bmatrix} 1 \\ -1 \end{bmatrix} + \tfrac{1}{3} e^{4t} \begin{bmatrix} 4 \\ -1 \end{bmatrix}.$$

Example 3.1

Solve the IVP $\dot{\mathbf{x}} = \begin{bmatrix} 17 & -6 \\ 45 & -16 \end{bmatrix} \mathbf{x}, \ \mathbf{x}(0) = \begin{bmatrix} 1 \\ 2 \end{bmatrix}$.

$$|A - \lambda I| = \begin{vmatrix} 17-\lambda & -6 \\ 45 & -16-\lambda \end{vmatrix} = (17-\lambda)(-16-\lambda) + 270 \ .$$

Therefore $\lambda^2 - \lambda - 2 = (\lambda - 2)(\lambda + 1) = 0$ for eigenvalues .

Therefore $\lambda_1 = 2, \ \lambda_2 = -1$.

For $\lambda_1 = 2$,

$$(A - \lambda_1 I)\mathbf{v}_1 = \begin{bmatrix} 15 & -6 \\ 45 & -18 \end{bmatrix} \mathbf{v}_1 = \mathbf{0} \text{ for eigenvectors} \ .$$

Therefore $\mathbf{v}_1 = \begin{bmatrix} 2 \\ 5 \end{bmatrix}$.

For $\lambda_2 = -1$,

$$(A - \lambda_2 I)\mathbf{v}_2 = \begin{bmatrix} 18 & -6 \\ 45 & -15 \end{bmatrix} \mathbf{v}_2 = \mathbf{0} \text{ for eigenvectors .}$$

Therefore $\mathbf{v}_2 = \begin{bmatrix} 1 \\ 3 \end{bmatrix}$.

The general solution is $\mathbf{x}(t) = c_1 e^{2t} \begin{bmatrix} 2 \\ 5 \end{bmatrix} + c_2 e^{-t} \begin{bmatrix} 1 \\ 3 \end{bmatrix}$.

The initial condition is $\mathbf{x}(0) = \begin{bmatrix} 1 \\ 2 \end{bmatrix}$. Putting $t = 0$ in the general solution gives

$$\mathbf{x}(0) = c_1 e^0 \begin{bmatrix} 2 \\ 5 \end{bmatrix} + c_2 e^0 \begin{bmatrix} 1 \\ 3 \end{bmatrix} = \begin{bmatrix} 1 \\ 2 \end{bmatrix} ,$$

i.e. $2c_1 + c_2 = 1$

$5c_1 + 3c_2 = 2$.

Therefore $c_1 = 1, \ c_2 = -1$.

Therefore $\mathbf{x} = e^{2t} \begin{bmatrix} 2 \\ 5 \end{bmatrix} - e^{-t} \begin{bmatrix} 1 \\ 3 \end{bmatrix}$ is the solution to the IVP.

Example 3.2

Show that $\mathbf{v}_1 = \begin{bmatrix} 1 \\ 1 \\ 1 \end{bmatrix}$ and $\mathbf{v}_2 = \begin{bmatrix} 1 \\ 2 \\ 4 \end{bmatrix}$ are eigenvectors of the matrix

$$A = \begin{bmatrix} 0 & 1 & 0 \\ 0 & 0 & 1 \\ 4 & -8 & 5 \end{bmatrix} .$$

Hence solve the initial value problem

$$\dot{x}_1 = x_2$$

$$\dot{x}_2 = x_3$$

$$\dot{x}_3 = 4x_1 - 8x_2 + 5x_3 \ ,$$

with $\quad x_1(0) = 2, \ x_2(0) = 3, \ x_3(0) = 5$

$$A\mathbf{v}_1 = \begin{bmatrix} 0 & 1 & 0 \\ 0 & 0 & 1 \\ 4 & -8 & 5 \end{bmatrix} \begin{bmatrix} 1 \\ 1 \\ 1 \end{bmatrix} = \begin{bmatrix} 1 \\ 1 \\ 1 \end{bmatrix} = \lambda_1 \mathbf{v}_1 \text{ with } \lambda_1 = 1 \ .$$

Therefore $\mathbf{v}_1$ is an eigenvector corresponding to eigenvalue $\lambda_1 = 1$.

$$A\mathbf{v}_2 = \begin{bmatrix} 0 & 1 & 0 \\ 0 & 0 & 1 \\ 4 & -8 & 5 \end{bmatrix} \begin{bmatrix} 1 \\ 2 \\ 4 \end{bmatrix} = \begin{bmatrix} 2 \\ 4 \\ 8 \end{bmatrix} = \lambda_2 \mathbf{v}_2 \text{ with } \lambda_2 = 2 \ .$$

Therefore $\mathbf{v}_2$ is an eigenvector corresponding to eigenvalue $\lambda_2 = 2$.
Now $\lambda_1 + \lambda_2 + \lambda_3 = \text{Trace}(A) = 5$. (See Problem 6, Section 2.2). Therefore $\lambda_3 = 2$ and we have a repeated eigenvalue.

For $\lambda_2 = 2$,

$$(A - \lambda_2 I) = \begin{bmatrix} -2 & 1 & 0 \\ 0 & -2 & 1 \\ 4 & -8 & 3 \end{bmatrix}$$

which has degeneracy equal to 1, and there is only one independent eigenvector corresponding to $\lambda_2 = 2$. Therefore to solve $\dot{\mathbf{x}} = A\mathbf{x}$ we must find a vector $\mathbf{w}_2$ such that

$$(A - \lambda_2 I)^2 \mathbf{w}_2 = \mathbf{0} \ ,$$

$$(A - \lambda_2 I)\mathbf{w}_2 \neq \mathbf{0} \ .$$

$(A - 2I)^2 \mathbf{w}_2 =$

$$\begin{bmatrix} -2 & 1 & 0 \\ 0 & -2 & 1 \\ 4 & -8 & 3 \end{bmatrix} \begin{bmatrix} -2 & 1 & 0 \\ 0 & -2 & 1 \\ 4 & -8 & 3 \end{bmatrix} \mathbf{w}_2 = \begin{bmatrix} 4 & -4 & 1 \\ 4 & -4 & 1 \\ 4 & -4 & 1 \end{bmatrix} \mathbf{w}_2$$

which will be $\mathbf{0}$ when $\mathbf{w}_2 = \begin{bmatrix} 1 \\ 1 \\ 0 \end{bmatrix}$.

Also $(A - 2I)\mathbf{w}_2 = \begin{bmatrix} -2 & 1 & 0 \\ 0 & -2 & 1 \\ 4 & -8 & 3 \end{bmatrix} \begin{bmatrix} 1 \\ 1 \\ 0 \end{bmatrix} = \begin{bmatrix} -1 \\ -2 \\ -4 \end{bmatrix} \neq \mathbf{0}$.

Therefore $\mathbf{x} = e^t \begin{bmatrix} 1 \\ 1 \\ 1 \end{bmatrix}$, $\mathbf{x} = e^{2t} \begin{bmatrix} 1 \\ 2 \\ 4 \end{bmatrix}$ and

$\mathbf{x} = e^{2t} \left\{ \begin{bmatrix} 1 \\ 1 \\ 0 \end{bmatrix} + \begin{bmatrix} -1 \\ -2 \\ -4 \end{bmatrix} t \right\}$ form a basis for the set of solutions of

$\dot{\mathbf{x}} = A\mathbf{x}$. Therefore the general solution is

$$\mathbf{x} = c_1 e^t \begin{bmatrix} 1 \\ 1 \\ 1 \end{bmatrix} + c_2 e^{2t} \begin{bmatrix} 1 \\ 2 \\ 4 \end{bmatrix} + c_3 e^{2t} \begin{bmatrix} 1-t \\ 1-2t \\ -4t \end{bmatrix} .$$

The initial condition is $\mathbf{x} = \begin{bmatrix} 2 \\ 3 \\ 5 \end{bmatrix}$ when $t = 0$. Therefore

$$\begin{bmatrix} 2 \\ 3 \\ 5 \end{bmatrix} = \begin{bmatrix} c_1 \\ c_1 \\ c_1 \end{bmatrix} + \begin{bmatrix} c_2 \\ 2c_2 \\ 4c_2 \end{bmatrix} + \begin{bmatrix} c_3 \\ c_3 \\ 0 \end{bmatrix} ,$$

i.e. $2 = c_1 + c_2 + c_3$

$3 = c_1 + 2c_2 + c_3$

$5 = c_1 + 4c_2$.

Therefore $c_1 = 1,\ c_2 = 1,\ c_3 = 0$.

Therefore $\mathbf{x} = e^t \begin{bmatrix} 1 \\ 1 \\ 1 \end{bmatrix} + e^{2t} \begin{bmatrix} 1 \\ 2 \\ 4 \end{bmatrix}$ is the solution of the IVP,

i.e. $x_1 = e^t + e^{2t}$

$x_2 = e^t + 2e^{2t}$

$x_3 = e^t + 4e^{2t}$.

Problems

1.

Solve $\dot{x} = \begin{bmatrix} 16 & -8 \\ -12 & 12 \end{bmatrix} \mathbf{x},\ \mathbf{x}(0) = \begin{bmatrix} 2 \\ 3 \end{bmatrix}$.

2. Solve $\dot{x} = \begin{bmatrix} 0 & 1 \\ -2 & -2 \end{bmatrix} \mathbf{x},\ \mathbf{x}(0) = \begin{bmatrix} 1 \\ -1 \end{bmatrix}$.

3. Solve $\dot{\mathbf{x}} = \begin{bmatrix} 0 & 1 \\ -9 & 6 \end{bmatrix} \mathbf{x},\ \mathbf{x}(0) = \begin{bmatrix} 1 \\ 4 \end{bmatrix}$.

4.

Solve $\dot{\mathbf{x}} = \begin{bmatrix} 0 & 1 & 0 \\ 0 & 0 & 1 \\ 2 & 1 & -2 \end{bmatrix} \mathbf{x},\ \mathbf{x}(0) = \begin{bmatrix} 1 \\ 0 \\ 1 \end{bmatrix}$.

5.

Solve $\dot{\mathbf{x}} = \begin{bmatrix} 0 & 1 & 1 \\ 0 & -1 & 1 \\ 0 & 0 & -1 \end{bmatrix} \mathbf{x},\ \mathbf{x}(0) = \begin{bmatrix} 2 \\ 0 \\ -1 \end{bmatrix}$.

6.

Solve $\dot{\mathbf{x}} = \begin{bmatrix} 1 & 1 & 0 \\ 0 & 1 & 0 \\ 0 & 1 & 1 \end{bmatrix} \mathbf{x},\ \mathbf{x}(0) = \begin{bmatrix} 1 \\ 1 \\ 1 \end{bmatrix}$.

7.

$$\text{Solve } \dot{\mathbf{x}} = \begin{bmatrix} -1 & 1 & 0 \\ 0 & -1 & 1 \\ 0 & 0 & -1 \end{bmatrix} \mathbf{x}, \ \mathbf{x}(0) = \begin{bmatrix} 1 \\ 1 \\ 2 \end{bmatrix}.$$

8.

$$\text{Solve } \dot{\mathbf{x}} = \begin{bmatrix} 1 & 1 & 0 \\ 0 & 1 & 0 \\ 0 & 0 & -2 \end{bmatrix} \mathbf{x}, \ \mathbf{x}(0) = \begin{bmatrix} 1 \\ 0 \\ 1 \end{bmatrix}.$$

Answers to problems

1.

$$\mathbf{x} = e^{4t} \begin{bmatrix} 2 \\ 3 \end{bmatrix}$$

2.

$$\mathbf{x} = \begin{bmatrix} e^{-t} \cos t \\ -e^{-t} (\cos t + \sin t) \end{bmatrix}$$

3.

$$\mathbf{x} = e^{3t} \begin{bmatrix} 1+t \\ 4+3t \end{bmatrix}$$

4.

$$\mathbf{x} = \tfrac{1}{2}e^{t} \begin{bmatrix} 1 \\ 1 \\ 1 \end{bmatrix} + \tfrac{1}{2}e^{-t} \begin{bmatrix} 1 \\ -1 \\ 1 \end{bmatrix} = \begin{bmatrix} \cosh t \\ \sinh t \\ \cosh t \end{bmatrix}$$

5.

$$\mathbf{x} = e^{-t} \begin{bmatrix} 2+t \\ -t \\ -1 \end{bmatrix}$$

6.

$$\mathbf{x} = e^{t} \begin{bmatrix} 1+t \\ 1 \\ 1+t \end{bmatrix}$$

7.

$$\mathbf{x} = e^{-t} \begin{bmatrix} 1 + t + t^2 \\ 1 + 2t \\ 2 \end{bmatrix}$$

8.

$$\mathbf{x} = e^{t} \begin{bmatrix} 1 \\ 0 \\ 0 \end{bmatrix} + e^{-2t} \begin{bmatrix} 0 \\ 0 \\ 1 \end{bmatrix} .$$

3.3 GEOMETRICAL INTERPRETATION OF THE SOLUTION OF $\dot{\mathbf{x}} = A\mathbf{x}$

We consider in particular the two-dimensional case. The solution of the initial value problem.

$$\begin{bmatrix} \dot{x}_1 \\ \dot{x}_2 \end{bmatrix} = A \begin{bmatrix} x_1 \\ x_2 \end{bmatrix} \text{, with } \begin{bmatrix} x_1(0) \\ x_2(0) \end{bmatrix} \text{ given ,}$$

can be represented by a curve in the x_1x_2 plane starting at the point $(x_1(0), x_2(0))$, see Fig. 3.1.

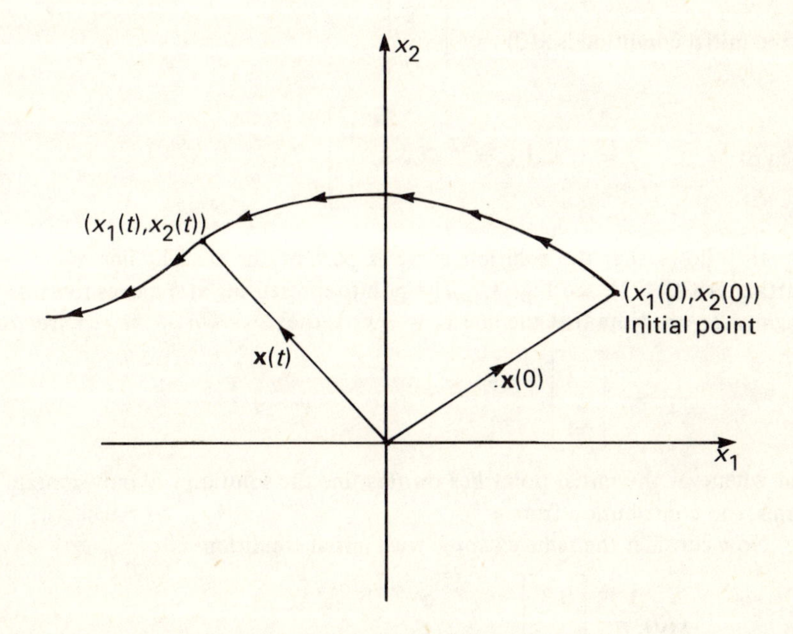

Fig. 3.1

The point $(x_1(t), x_2(t))$ moves along the SOLUTION CURVE as t increases. Alternatively we can think of the solution curve being swept out by the terminal point of the vector.

$$\mathbf{x}(t) = \begin{bmatrix} x_1 \\ x_2 \end{bmatrix}, \quad \mathbf{x}(t) \text{ drawn } \textit{from origin} .$$

Consider the example $\dot{\mathbf{x}} = \begin{bmatrix} 0 & 1 \\ 2 & 1 \end{bmatrix} \mathbf{x}$. The eigenvalues are $\lambda_1 = -1$, $\lambda_2 = 2$ with corresponding eigenvectors

$$\mathbf{v}_1 = \begin{bmatrix} 1 \\ -1 \end{bmatrix}, \quad \mathbf{v}_2 = \begin{bmatrix} 1 \\ 2 \end{bmatrix} .$$

Therefore the general solution is

$$\mathbf{x} = c_1 e^{-t} \begin{bmatrix} 1 \\ -1 \end{bmatrix} + c_2 e^{2t} \begin{bmatrix} 1 \\ 2 \end{bmatrix} .$$

If the initial condition is $\mathbf{x}(0) = \begin{bmatrix} 2 \\ -2 \end{bmatrix}$,

then $c_1 = 2$, $c_2 = 0$ and $\mathbf{x} = 2e^{-t} \begin{bmatrix} 1 \\ -1 \end{bmatrix}$.

It follows that the solution curve is part of the straight line $x_2 = -x_1$ starting at $(2, -2)$, see Fig. 3.2. The point representing $\mathbf{x}(t)$ moves towards the origin as $t \to \infty$. Note that the line $x_2 = -x_1$ is the *direction of the eigenvector*

$$\mathbf{v}_1 = \begin{bmatrix} 1 \\ -1 \end{bmatrix},$$

and whenever the initial point lies on this line the solution will only contain e^{-t} terms, (no contribution from e^{2t}).

Now consider the same example with initial condition

$$\mathbf{x}(0) = \begin{bmatrix} 0.3 \\ 0.6 \end{bmatrix} .$$

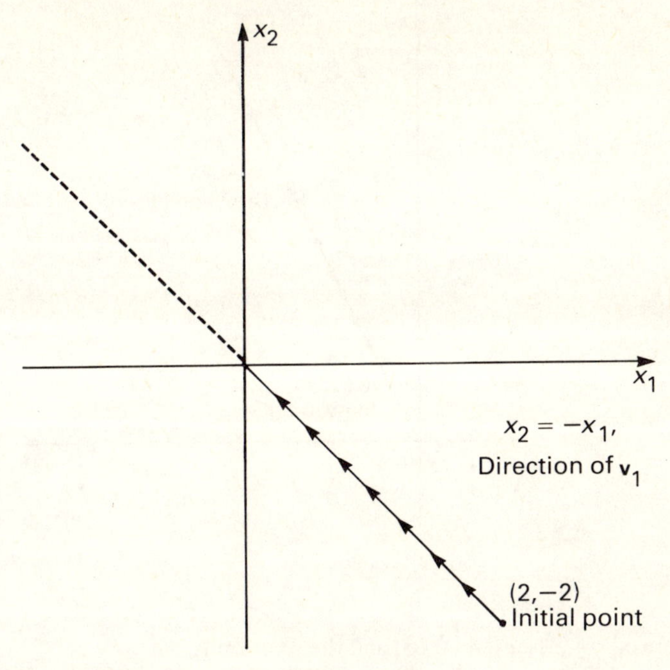

Fig. 3.2

Here $c_1 = 0, c_2 = 0.3$ and so $\mathbf{x} = 0.3\, e^{2t} \begin{bmatrix} 1 \\ 2 \end{bmatrix}$.

The solution curve is now part of the line $x_2 = 2x_1$ starting at (0.3, 0.6), see Fig. 3.3. The point representing $\mathbf{x}(t)$ moves away from the origin as $t \to \infty$. Note that the line $x_2 = 2x_1$ is the direction of the eigenvector

$$\mathbf{v}_2 = \begin{bmatrix} 1 \\ 2 \end{bmatrix},$$

and whenever the initial point lies on this line the solution will contain only e^{2t} terms.

Finally consider a situation when the initial point does not lie on either of the eigenvector directions,

$$\mathbf{x}(0) = \begin{bmatrix} 2.15 \\ -2 \end{bmatrix}.$$

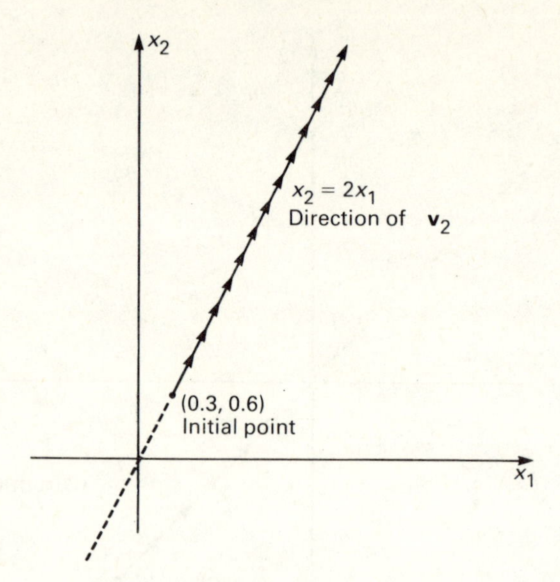

Fig. 3.3

Therefore $c_1 = 2.1$, $c_2 = 0.05$, and the solution is

$$\mathbf{x} = 2.1\, \mathrm{e}^{-t} \begin{bmatrix} 1 \\ -1 \end{bmatrix} + 0.05\, \mathrm{e}^{2t} \begin{bmatrix} 1 \\ 2 \end{bmatrix} .$$

The solution contains both e^{-t} and e^{2t} terms, but as $t \to \infty$ the e^{2t} terms dominate. The point representing $\mathbf{x}$ ultimately moves away from the origin in the direction of the eigenvector $\mathbf{v}_2$, see Fig. 3.4.

Similar solution curves arise from other initial points on neither of the eigenvector directions. Figure 3.5 shows the family of solution curves to be a set of hyperbolae with the eigenvector directions as asymptotes. Note that all solutions, except those for which the initial point is on the line corresponding to eigenvector $\mathbf{v}_1$, tend to infinity in a direction $\pm\, \mathbf{v}_2$.

The above representation can be extended to three-dimensional cases by representing the solution of $\dot{\mathbf{x}} = A\mathbf{x}$ by a curve in $x_1\, x_2\, x_3$ space, see Fig. 3.6. Assuming that λ_1, λ_2 and λ_3 are distinct real eigenvalues of A with corresponding eigenvectors $\mathbf{v}_1, \mathbf{v}_2, \mathbf{v}_3$, the solution is

$$\mathbf{x} = c_1 \mathrm{e}^{\lambda_1 t}\mathbf{v}_1 + c_2 \mathrm{e}^{\lambda_2 t}\mathbf{v}_2 + c_3 \mathrm{e}^{\lambda_3 t}\mathbf{v}_3 \quad .$$

We can choose the initial point $(x_1(0), x_2(0), x_3(0))$ to exclude some of the terms from the general solution. For example, if the initial point is in the direction of $\mathbf{v}_1$, then the solution curve will be a straight line in this direction ($\mathbf{x}$ will contain only terms in $\mathrm{e}^{\lambda_1 t}$).

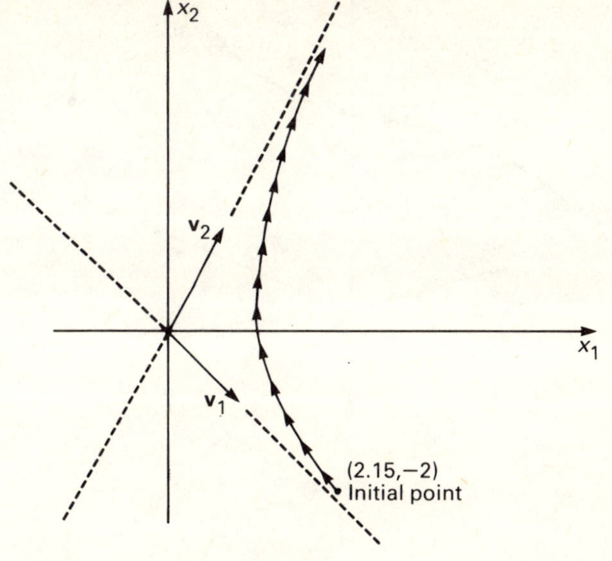

Fig. 3.4

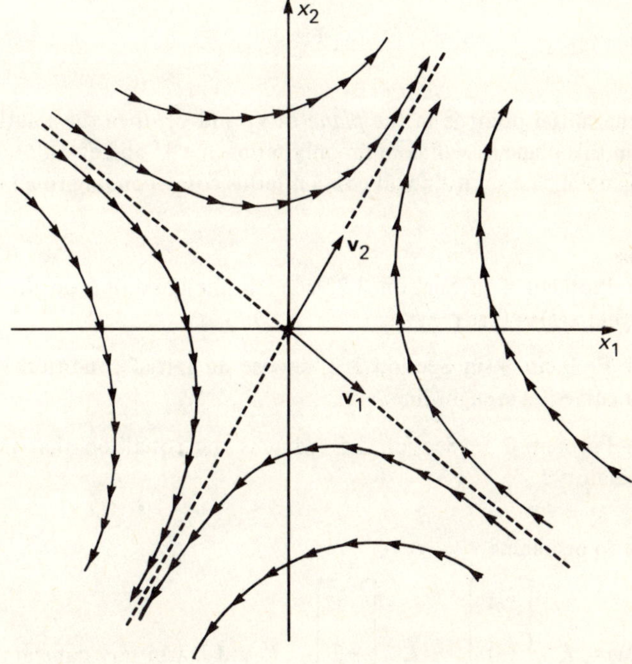

Fig. 3.5 – Solution curves for $\dot{\mathbf{x}} = \begin{bmatrix} 0 & 1 \\ 2 & 1 \end{bmatrix} \mathbf{x}$.

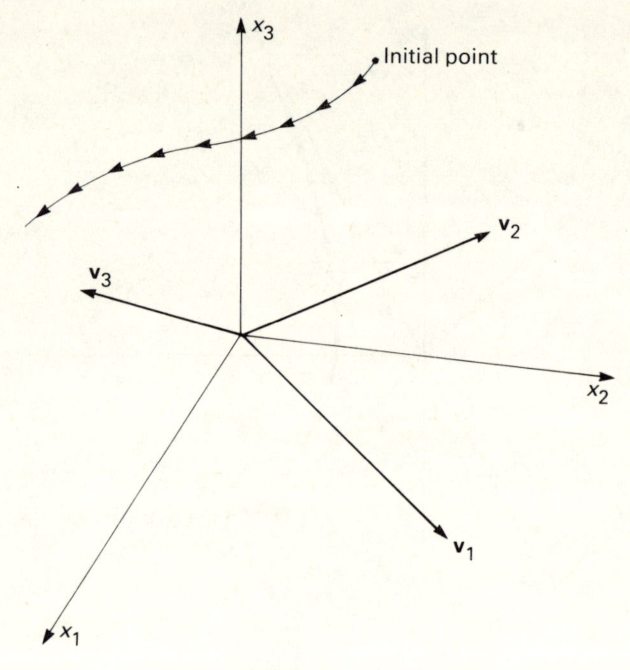

Fig. 3.6

If the initial point is in the *plane* of $\mathbf{v}_1$ and $\mathbf{v}_2$, then the solution curve will remain in this plane ($\mathbf{x}$ will contain only terms in $e^{\lambda_1 t}$ and $e^{\lambda_2 t}$).

In general, the solution will contain terms corresponding to all eigenvalues.

Problems

1. For Problem 4 in Section 3.2, what is the condition on the initial vector $\mathbf{x}(0)$ so that $\mathbf{x}(t) \to 0$ as $t \to \infty$?

2. For Problem 7 in Section 3.2, choose an initial condition such that the solution curve is a straight line.

3. For Problem 2 in Section 2.5, what is the condition that the solution is non-oscillatory?

Answers to problems

1.

$$\mathbf{x}(0) = K \begin{bmatrix} 1 \\ -1 \\ 1 \end{bmatrix} + L \begin{bmatrix} 1 \\ -2 \\ 4 \end{bmatrix}, K \text{ and } L \text{ arbitrary constants}.$$

2.

$$\mathbf{x}(0) = \begin{bmatrix} K \\ 0 \\ 0 \end{bmatrix} \quad \text{where } K \text{ is an arbitrary constant .}$$

3.

$$\mathbf{x}(0) = K \begin{bmatrix} 1 \\ -1 \\ 1 \end{bmatrix} \quad \text{where } K \text{ is an arbitrary constant .}$$

3.4 STABILITY AND THE NATURE OF SOLUTION CURVES

The system of differential equations $\dot{\mathbf{x}} = A\mathbf{x}$ has a trivial solution $\mathbf{x} = \mathbf{0}$. This zero solution is represented by the origin of coordinates in the solution space and the origin is called an EQUILIBRIUM POINT since $\dot{\mathbf{x}} = \mathbf{0}$ when $\mathbf{x} = \mathbf{0}$. Further if the determinant of the matrix A is non-zero then the origin is the *only* equilibrium point.

We define the origin to be a STABLE equilibrium point for the system $\dot{\mathbf{x}} = A\mathbf{x}$ if $\mathbf{x} \to \mathbf{0}$ as $t \to \infty$ for *all* initial points. This will be the case if and only if all the eigenvalues of A are negative real numbers or complex numbers with negative real parts. If only one of the eigenvalues is positive (or complex with positive real part), then $\mathbf{x} \to \infty$ as $t \to \infty$ unless the initial point $\mathbf{x}(0)$ is chosen to exclude the eigenvalue. In such a case the solution tends to $\mathbf{0}$ (theoretically) but the equilibrium is UNSTABLE, as the slightest disturbance (always present in a practical situation) will introduce the positive eigenvalue term and $\mathbf{x}$ will tend to ∞. [Unstable equilibrium is analogous to balancing one snooker ball on top of another.]

Stability (and instability) may be either OSCILLATORY or NON-OSCILLATORY depending on whether or not sine and cosine terms are present in the solution of $\dot{\mathbf{x}} = A\mathbf{x}$, that is, depending on the presence or absence of complex conjugate eigenvalues. We shall consider, in particular, the two-dimensional system

$$\begin{bmatrix} \dot{x}_1 \\ \dot{x}_2 \end{bmatrix} = A \begin{bmatrix} x_1 \\ x_2 \end{bmatrix}$$

and distinguish four types of equilibrium point as follows.

(i) The equilibrium point is called a SADDLE if the eigenvalues of A are *real and of opposite sign*, i.e. $\mathbf{x} \to \infty$ as $t \to \infty$. See Fig. 3.7 and compare with Fig. 3.5. A saddle is always an *unstable* equilibrium point.

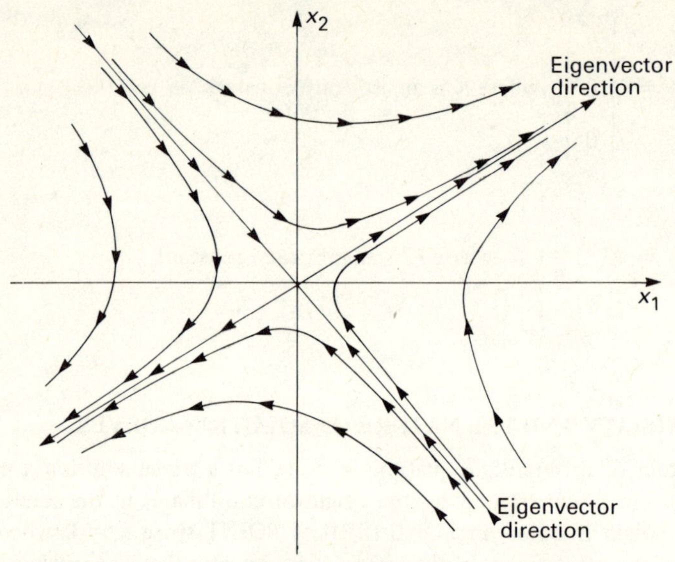

Fig. 3.7 – Saddle.

(ii) The equilibrium point is called a NODE if the eigenvalues of A are *real and of the same sign.* If both are negative then the node is stable, i.e. $\mathbf{x} \to \mathbf{0}$ without oscillation as $t \to \infty$. If both eigenvalues are positive the node is unstable, $\mathbf{x} \to \infty$ without oscillation as $t \to \infty$. Figure 3.8 shows a stable node.

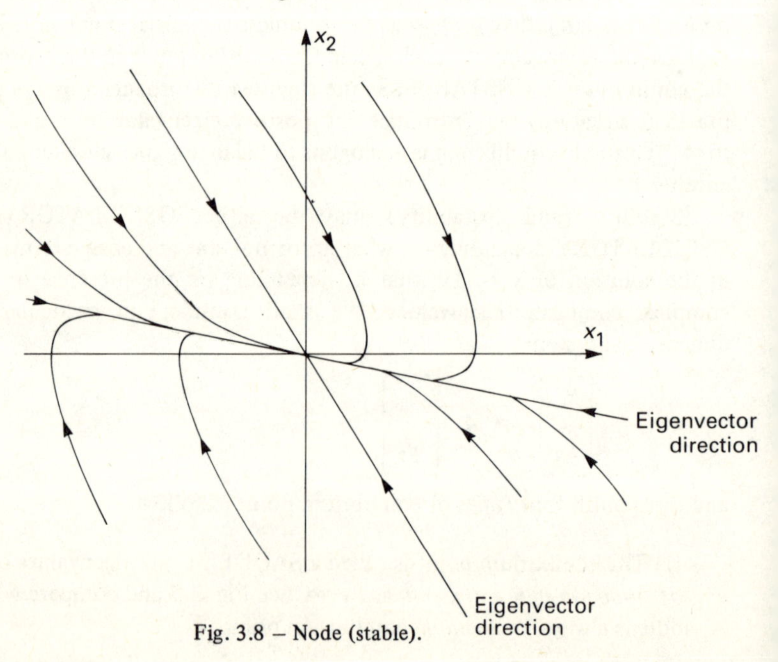

Fig. 3.8 – Node (stable).

(iii) The equilibrium point is called a FOCUS if the eigenvalues of A are *complex conjugates,* $\alpha \pm j\beta$, $\alpha \neq 0$. The solution is oscillatory and stable for $\alpha < 0$, unstable for $\alpha > 0$. Figure 3.9 shows a stable focus, the solution curves spiralling around the origin.

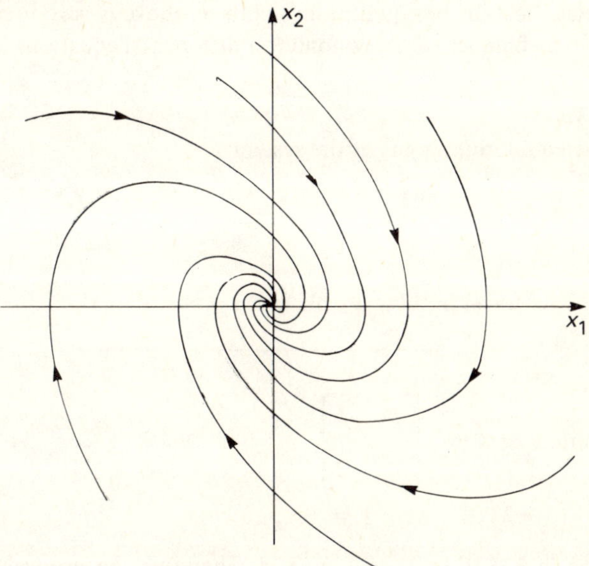

Fig. 3.9 – Focus (stable).

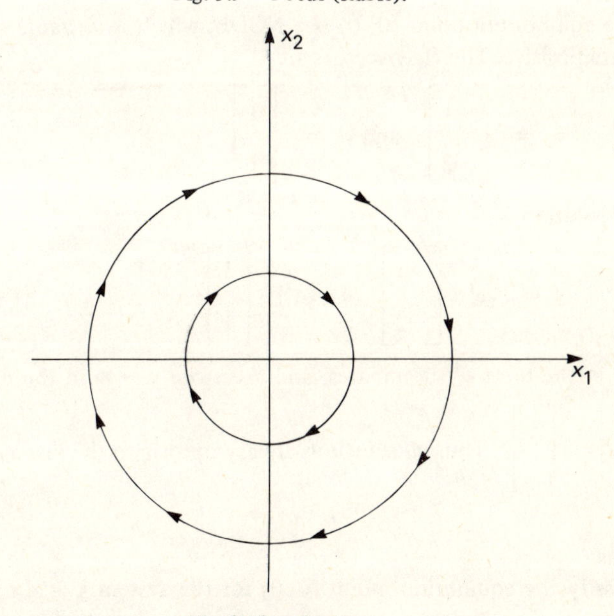

Fig. 3.10 – Centre.

(iv) The equilibrium point is called a CENTRE if the eigenvalues of A are *purely imaginary*. The solution is periodic and the solution curves are closed. Figure 3.10 shows a centre; the solution is oscillatory but neither stable nor unstable.

The classification of equilibrium points as above is very useful when using linearisation techniques to study nonlinear differential equations, see Chapter 8.

Example 3.3

Classify the equilibrium point of the system

$$\dot{\mathbf{x}} = \begin{bmatrix} 3 & 1 \\ 2 & 2 \end{bmatrix} \mathbf{x} \ .$$

Describe the solutions in the x_1x_2 plane as $t \to \infty$.

The eigenvalues of $A = \begin{bmatrix} 3 & 1 \\ 2 & 2 \end{bmatrix}$ are solutions of

$$(3-\lambda)(2-\lambda)-2 = 0 \ ,$$

i.e. $\lambda^2 - 5\lambda + 4 = 0$, so $\lambda_1 = 1$, $\lambda_2 = 4$. Therefore the eigenvalues are real and of the same sign.

The equilibrium point (0, 0) is a NODE, which is *unstable* since the eigenvalues are positive. The eigenvectors are

$$\mathbf{v}_1 = \begin{bmatrix} 1 \\ -2 \end{bmatrix} \quad \text{and } \mathbf{v}_2 = \begin{bmatrix} 1 \\ 1 \end{bmatrix}$$

and the solution is

$$\mathbf{x} = c_1 e^t \begin{bmatrix} 1 \\ -2 \end{bmatrix} + c_2 e^{4t} \begin{bmatrix} 1 \\ 1 \end{bmatrix} .$$

As $t \to \infty$ the term e^{4t} dominates, and therefore $x \to \infty$ in the direction of the vector $\mathbf{v}_2 = \begin{bmatrix} 1 \\ 1 \end{bmatrix}$. Thus *all* solutions are asymptotic to the line $x_2 = x_1$.

Problems

1. Classify the equilibrium point (0, 0) for the system $\dot{\mathbf{x}} = A\mathbf{x}$ for each of the following matrices. Indicate the stability where appropriate.

(i) $\begin{bmatrix} 0 & -2 \\ 2 & 0 \end{bmatrix}$, (ii) $\begin{bmatrix} -4 & 1 \\ 1 & -2 \end{bmatrix}$,

(iii) $\begin{bmatrix} -4 & 3 \\ 5 & -2 \end{bmatrix}$, (iv) $\begin{bmatrix} 4 & 1 \\ -1 & 2 \end{bmatrix}$,

(v) $\begin{bmatrix} -6 & -3 \\ 6 & 2 \end{bmatrix}$.

2. For the system $\begin{bmatrix} \dot{x}_1 \\ \dot{x}_2 \end{bmatrix} = \begin{bmatrix} a & b \\ c & d \end{bmatrix} \begin{bmatrix} x_1 \\ x_2 \end{bmatrix}$

prove the following conditions for classification of the equilibrium point (0, 0).

(i) Saddle if $ad - bc < 0$.

(ii) Stable node if $(a+d)^2 \geqslant 4(ad-bc) > 0$ *and* $a+d < 0$.

(iii) Unstable node if $(a+d)^2 \geqslant 4(ad-bc) > 0$ *and* $a+d > 0$.

(iv) Stable focus if $(a+d)^2 < 4(ad-bc)$ *and* $a+d < 0$.

(v) Unstable focus if $(a+d)^2 < 4(ad-bc)$ *and* $a+d > 0$.

(vi) Centre if $a+d = 0$ *and* $ad-bc > 0$.

3. Given that $\lambda_1 = \lambda_2 > 0$ for a 2×2 matrix A with only one independent eigenvector $\mathbf{v}_1$, show that as $t \to \infty$ all solution curves in the x_1x_2 plane are asymptotic to a line through (0,0) parallel to the vector $\mathbf{v}_1$.

Answers to problems

1. (i) Centre.
   (ii) Stable node.
   (iii) Saddle, unstable.
   (iv) Unstable node.
   (v) Stable focus.

4

Extensions to higher order

4.1 INTRODUCTION

In this chapter we deal with certain systems of differential equations which may be solved by reducing them to the form $\dot{\mathbf{x}} = A\mathbf{x}$. Two situations are considered:

(i) linear systems with constant coefficients involving higher derivatives than first order,
(ii) forced systems of the form $\dot{\mathbf{x}} = A\mathbf{x} + \mathbf{f}(t)$ which can be reduced to homogeneous form provided the components of the vector $\mathbf{f}(t)$ belong to a certain class of functions.

As a particular case of (i) we consider *undamped* systems of second order equations, which may be solved using eigenvalue techniques without first being reduced to first order form.

4.2 REDUCTION OF HIGHER ORDER SYSTEMS TO $\dot{\mathbf{x}} = A\mathbf{x}$

The method for reduction of systems involving derivatives higher than first order is explained in the following examples.

Example 4.1
Reduce the system

$$\ddot{x}_1 = 2x_1 + \dot{x}_1 + x_2$$

$$\dot{x}_2 = 4x_1 + 2x_2$$

to first order form and complete the solution.

Let
$$y_1 = x_1$$

$$y_2 = \dot{x}_1$$

$$y_3 = x_2$$

Therefore
$$\begin{aligned}\dot{y}_1 &= \dot{x}_1 = y_2\\ \dot{y}_2 &= \ddot{x}_1 = 2x_1 + \dot{x}_1 + x_2 = 2y_1 + y_2 + y_3\\ \dot{y}_3 &= \dot{x}_2 = 4x_1 + 2x_2 = 4y_1 + 2y_3 \quad .\end{aligned}$$

Thus we have a three-dimensional system of the form $\dot{\mathbf{y}} = A\mathbf{y}$,

i.e.
$$\begin{bmatrix}\dot{y}_1\\ \dot{y}_2\\ \dot{y}_3\end{bmatrix} = \begin{bmatrix}0 & 1 & 0\\ 2 & 1 & 1\\ 4 & 0 & 2\end{bmatrix}\begin{bmatrix}y_1\\ y_2\\ y_3\end{bmatrix} .$$

The eigenvalues of the matrix A are $\lambda_1 = 3,\ \lambda_2 = 0,\ \lambda_3 = 0$ with eigenvectors

$$\mathbf{v}_1 = \begin{bmatrix}1\\ 3\\ 4\end{bmatrix},\quad \mathbf{v}_2 = \begin{bmatrix}1\\ 0\\ -2\end{bmatrix}$$

(only one independent eigenvector corresponding to $\lambda_2 = 0$).

$$\text{For } \lambda_2 = 0,\ (A - \lambda_2 I)^2\mathbf{w}_2 = \begin{bmatrix}2 & 1 & 1\\ 6 & 3 & 3\\ 8 & 4 & 4\end{bmatrix}\mathbf{w}_2 = \mathbf{0} .$$

$$\text{Take } \mathbf{w}_2 = \begin{bmatrix}0\\ 1\\ -1\end{bmatrix} \text{ and } (A - \lambda_2 I)\mathbf{w}_2 = \begin{bmatrix}1\\ 0\\ -2\end{bmatrix} .$$

$$\text{Therefore } \mathbf{w}_2 + (A - \lambda_2 I)\mathbf{w}_2 t = \begin{bmatrix}t\\ 1\\ -1-2t\end{bmatrix} .$$

The general solution of $\dot{\mathbf{y}} = A\mathbf{y}$ is

$$\begin{bmatrix}y_1\\ y_2\\ y_3\end{bmatrix} = c_1 e^{3t}\begin{bmatrix}1\\ 3\\ 4\end{bmatrix} + c_2\begin{bmatrix}1\\ 0\\ -2\end{bmatrix} + c_3\begin{bmatrix}t\\ 1\\ -1-2t\end{bmatrix} .$$

However, $x_1 = y_1$ and $x_2 = y_3$. Therefore the solution of the given system is

$$x_1 = c_1 e^{3t} + c_2 + c_3 t$$
$$x_2 = 4c_1 e^{3t} - 2c_2 - c_3(1+2t) \ .$$

Example 4.2
Solve the differential equation

$$\frac{d^4 y}{dt^4} - y = 0$$

by reducing it to the form $\dot{\mathbf{x}} = A\mathbf{x}$.

Let
$$x_1 = y$$
$$x_2 = \dot{y}$$
$$x_3 = \ddot{y}$$
$$x_4 = \dddot{y} \ .$$

Therefore
$$\dot{x}_1 = \dot{y} = x_2$$
$$\dot{x}_2 = \ddot{y} = x_3$$
$$\dot{x}_3 = \dddot{y} = x_4$$
$$\dot{x}_4 = \ddddot{y} = y = x_1 \ .$$

Thus we have the four-dimensional system

$$\begin{bmatrix} \dot{x}_1 \\ \dot{x}_2 \\ \dot{x}_3 \\ \dot{x}_4 \end{bmatrix} = \begin{bmatrix} 0 & 1 & 0 & 0 \\ 0 & 0 & 1 & 0 \\ 0 & 0 & 0 & 1 \\ 1 & 0 & 0 & 0 \end{bmatrix} \begin{bmatrix} x_1 \\ x_2 \\ x_3 \\ x_4 \end{bmatrix} .$$

The eigenvalues are $\lambda_1 = 1$, $\lambda_2 = -1$, $\lambda_3 = j$, $\lambda_4 = -j$ with eigenvectors

$$\mathbf{v}_1 = \begin{bmatrix} 1 \\ 1 \\ 1 \\ 1 \end{bmatrix}, \ \mathbf{v}_2 = \begin{bmatrix} 1 \\ -1 \\ 1 \\ -1 \end{bmatrix},$$

$$\mathbf{v}_3 = \begin{bmatrix} 1 \\ j \\ -1 \\ -j \end{bmatrix}, \quad \mathbf{v}_4 = \begin{bmatrix} 1 \\ -j \\ -1 \\ j \end{bmatrix}.$$

For $\lambda_3 = j$, $\mathbf{x} = e^{jt} \begin{bmatrix} 1 \\ j \\ -1 \\ -j \end{bmatrix}$ is a solution of $\dot{\mathbf{x}} = A\mathbf{x}$.

The real part $\begin{bmatrix} \cos t \\ -\sin t \\ -\cos t \\ \sin t \end{bmatrix}$

and the imaginary part $\begin{bmatrix} \sin t \\ \cos t \\ -\sin t \\ -\cos t \end{bmatrix}$ are also solutions .

Therefore the general solution of $\dot{\mathbf{x}} = A\mathbf{x}$ is

$$\begin{bmatrix} x_1 \\ x_2 \\ x_3 \\ x_4 \end{bmatrix} = c_1 e^t \begin{bmatrix} 1 \\ 1 \\ 1 \\ 1 \end{bmatrix} + c_2 e^{-t} \begin{bmatrix} 1 \\ -1 \\ 1 \\ -1 \end{bmatrix} + c_3 \begin{bmatrix} \cos t \\ -\sin t \\ -\cos t \\ \sin t \end{bmatrix} + c_4 \begin{bmatrix} \sin t \\ \cos t \\ -\sin t \\ -\cos t \end{bmatrix}.$$

However, $x_1 = y$ and the solution of the given equation is

$$y = c_1 e^t + c_2 e^{-t} + c_3 \cos t + c_4 \sin t.$$

Problems

1. Express the following equations in the form $\dot{\mathbf{x}} = A\mathbf{x}$ and complete the solution.

(i) $\ddot{y} + 4\dot{y} + 3y = 0$.

(ii) $\ddot{y} + 4\dot{y} + 4y = 0$.

(iii) $\dddot{y} - 8y = 0$.

(iv) $\ddddot{y} - 2\dddot{y} + 2\dot{y} - y = 0$.

2. Solve the following systems after reducing them to first order form.

(i) $\ddot{y} + 6\dot{y} + 8y = 8x$

$\dot{x} + 2x = 0$.

(ii) $\dot{x}_1 = x_2$

$\ddot{x}_2 = -\dot{x}_1 + 2\dot{x}_2$.

(iii) $\ddot{x}_1 = 2\dot{x}_1 - \dot{x}_2 + 2x_2$

$\ddot{x}_2 = \dot{x}_1 + 2x_2$.

Answers to problems

1. (i) $y = c_1 e^{-t} + c_2 e^{-3t}$.

(ii) $y = (c_1 + c_2 t)e^{-2t}$.

(iii) $y = c_1 e^{2t} + e^{-t}(c_2 \cos\sqrt{3}\, t + c_3 \sin\sqrt{3}\, t)$.

(iv) $y = c_1 e^{-t} + (c_2 + c_3 t + c_4 t^2)e^t$.

2. (i) $x = c_3 e^{-2t}$, $y = c_1 e^{-2t} + c_2 e^{-4t} + 4c_3 t e^{-2t}$.

(ii) $x_1 = c_1 + c_2 e^t + c_3 e^t(t-1)$, $x_2 = c_2 e^t + c_3 t e^t$.

(iii) $x_1 = c_1 + c_2 e^t + c_3 e^{-t} + c_4 e^{2t}$.

$x_2 = -c_2 e^t + c_3 e^{-t} + c_4 e^{2t}$.

4.3 UNDAMPED SECOND ORDER SYSTEMS

A system of second-order differential equations of the form,

$$\ddot{\mathbf{x}} = A\dot{\mathbf{x}} + B\mathbf{x} ,$$

may be solved using the method described in Section 4.2. If the matrix A is the zero matrix we have the UNDAMPED second order system,

$$\ddot{\mathbf{x}} = B\mathbf{x} \tag{4.1}$$

Undamped mechanical systems cannot exist in practice, but very lightly damped systems are common in the study of vibrations.

Vibrating systems are often analysed by *assuming* them to be undamped since a shorter method of solution is available for equation 4.1. We expect an undamped vibrating system to have a solution containing sine and cosine terms. Therefore we try a solution $\mathbf{x} = (\cos \omega t)\mathbf{v}$ [or $\mathbf{x} = (\sin \omega t)\mathbf{v}$] in equation 4.1, where $\mathbf{v}$ is a vector with constant components.

Substituting $\mathbf{x} = (\cos \omega t)\mathbf{v}$, $\ddot{\mathbf{x}} = -\omega^2(\cos \omega t)\mathbf{v}$ into equation 4.1 gives

$$-\omega^2\mathbf{v} = B\mathbf{v} \;,$$

i.e.

$$(B + \omega^2 I)\mathbf{v} = \mathbf{0} \;.$$

Therefore if we solve $(B - \lambda I)\,\mathbf{v} = \mathbf{0}$ to get the eigenvalues and eigenvectors of B, we can obtain possible values for ω from

$$\omega_1^2 = -\lambda_1, \; \omega_2^2 = -\lambda_2, \ldots \;.$$

For distinct eigenvalues of B, equation 4.1 has independent solutions

$$(\cos \omega_1 t)\mathbf{v}_1, \; (\sin \omega_1 t)\mathbf{v}_1, \; (\cos \omega_2 t)\mathbf{v}_2, \; (\sin \omega_2 t)\mathbf{v}_2, \ldots \;.$$

The general solution of equation 4.1 is then a linear combination of these solutions. The calculation of the values for ω involves finding the eigenvalues of the $n \times n$ matrix B, whereas the method of Section 4.2 requires working with a $2n \times 2n$ matrix.

Example 4.3

Solve the system

$$\ddot{\mathbf{x}} = \begin{bmatrix} -10 & 6 \\ 15 & -19 \end{bmatrix} \mathbf{x} \;.$$

$$|B - \lambda I| = (-10-\lambda)(-19-\lambda) - 90$$

$$= \lambda^2 + 29\lambda + 100 = 0 \text{ for eigenvalues} \;.$$

Therefore $\lambda_1 = -4, \; \lambda_2 = -25$.

Therefore $\omega_1^2 = 4, \; \omega_2^2 = 25$,

i.e. $\omega_1 = 2, \; \omega_2 = 5$.

The corresponding eigenvectors are obtained from

$$\begin{bmatrix} -6 & 6 \\ 15 & -15 \end{bmatrix} \mathbf{v}_1 = \mathbf{0} \text{ and } \begin{bmatrix} 15 & 6 \\ 15 & 6 \end{bmatrix} \mathbf{v}_2 = \mathbf{0} \;,$$

i.e. $\mathbf{v}_1 = \begin{bmatrix} 1 \\ 1 \end{bmatrix}, \mathbf{v}_2 = \begin{bmatrix} 2 \\ -5 \end{bmatrix}$.

Therefore the required solution is

$$\mathbf{x} = c_1 \cos 2t \begin{bmatrix} 1 \\ 1 \end{bmatrix} + c_2 \sin 2t \begin{bmatrix} 1 \\ 1 \end{bmatrix} + c_3 \cos 5t \begin{bmatrix} 2 \\ -5 \end{bmatrix} + c_4 \sin 5t \begin{bmatrix} 2 \\ -5 \end{bmatrix}$$

If Example 4.3 is attempted using the method of Section 4.2 we set up the vector

$$\begin{bmatrix} y_1 \\ y_2 \\ y_3 \\ y_4 \end{bmatrix} = \begin{bmatrix} x_1 \\ x_2 \\ \dot{x}_1 \\ \dot{x}_2 \end{bmatrix},$$

and then we obtain the first order system

$$\dot{\mathbf{y}} = \begin{bmatrix} 0 & 0 & 1 & 0 \\ 0 & 0 & 0 & 1 \\ -10 & 6 & 0 & 0 \\ 15 & -19 & 0 & 0 \end{bmatrix} \mathbf{y} .$$

The 4×4 matrix has eigenvalues $\pm 2j, \pm 5j$.

The reader is left to complete this solution and demonstrate its equivalence to the result given in Example 4.3.

Example 4.4

Consider the undamped mechanical system shown in Fig. 4.1. Assuming linear springs and neglecting friction in the wheels, derive the equations of motion and solve them for the case

$$\frac{k_1}{m_1} = 12, \quad \frac{k_2}{m_1} = \frac{k_2}{m_2} = 8 ,$$

$$y_1(0) = y_2(0) = 0, \ \dot{y}_1(0) = 0, \ \dot{y}_2(0) = 24 .$$

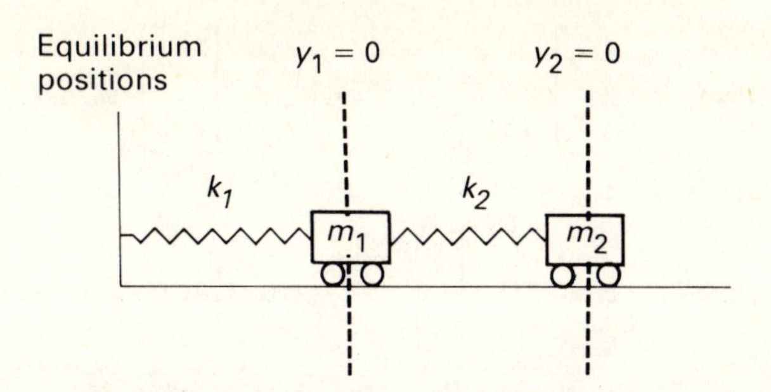

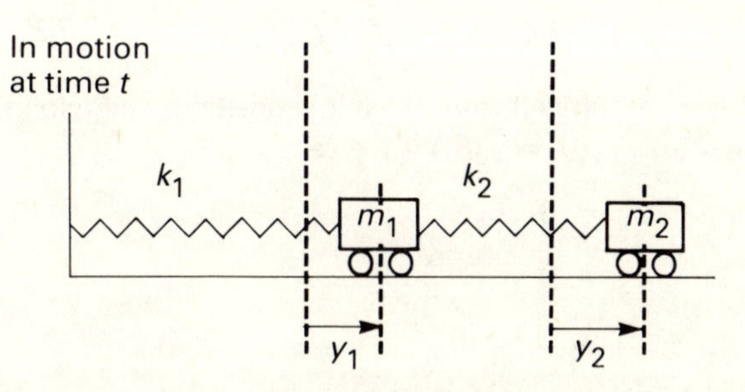

Fig. 4.1 – k_1 and k_2 are the spring stiffnesses, m_1 and m_2 are the masses of the trucks.

Applying Newton's second law to each mass gives

$$m_1\ddot{y}_1 = k_2(y_2 - y_1) - k_1y_1$$

$$m_2\ddot{y}_2 = -k_2(y_2 - y_1) \ ,$$

i.e.

$$\ddot{y}_1 = -\frac{(k_1 + k_2)}{m_1}y_1 + \frac{k_2}{m_1}\,y_2$$

$$\ddot{y}_2 = \frac{k_2}{m_2}\,y_1 - \frac{k_2}{m_2}\,y_2 \ .$$

Using the given numbers and writing in matrix form we obtain

$$\ddot{\mathbf{y}} = \begin{bmatrix} -20 & 8 \\ 8 & -8 \end{bmatrix} \mathbf{y} \ .$$

The eigenvalues and eigenvectors of

$$B = \begin{bmatrix} -20 & 8 \\ 8 & -8 \end{bmatrix} \quad \text{are}$$

$$\lambda_1 = -4, \mathbf{v}_1 = \begin{bmatrix} 1 \\ 2 \end{bmatrix}, \quad \lambda_2 = -24, \mathbf{v}_2 = \begin{bmatrix} 2 \\ -1 \end{bmatrix}.$$

Therefore $\omega_1 = 2, \ \omega_2 = \sqrt{24}$, and

$$\mathbf{y} = (c_1 \cos 2t + c_2 \sin 2t) \begin{bmatrix} 1 \\ 2 \end{bmatrix} + (c_3 \cos\sqrt{24}t + c_4 \sin\sqrt{24}t) \begin{bmatrix} 2 \\ -1 \end{bmatrix},$$

i.e. $y_1 = c_1 \cos 2t + c_2 \sin 2t + 2c_3 \cos\sqrt{24}t + 2c_4 \sin\sqrt{24}t$

$$y_2 = 2c_1 \cos 2t + 2c_2 \sin 2t - c_3 \cos\sqrt{24}t - c_4 \sin\sqrt{24}t \ .$$

c_1, c_2, c_3, c_4 are determined from the initial positions and velocities of the trucks.
The conditions $y_1(0) = y_2(0) = 0$ give

$$c_1 + 2c_3 = 0$$

$$2c_1 - c_3 = 0 \ ,$$

i.e. $\quad c_1 = c_3 = 0 \ .$

The condition $\dot{y}_1(0) = 0, \ \dot{y}_2(0) = 24$ give

$$2c_2 + 2\sqrt{24}\, c_4 = 0$$

$$4c_2 - \sqrt{24}\, c_4 = 24 \ ,$$

i.e. $c_2 = 4.8, \ c_4 = -\dfrac{\sqrt{24}}{5} \approx -0.98 \ .$

Therefore $y_1 = 4.8 \sin 2t - 1.96 \sin\sqrt{24}t$

$$y_2 = 9.6 \sin 2t + 0.98 \sin\sqrt{24}t \ .$$

Note that the solution in Example 4.4 consists of a combination of terms of frequency 2 radians/second and terms of frequency $\sqrt{24}$ radians/second. By choosing suitable initial conditions we get a solution exhibiting one frequency only. In particular the conditions

$$\mathbf{y}(0) = \begin{bmatrix} 1 \\ 2 \end{bmatrix} = \mathbf{v}_1, \ \dot{\mathbf{y}}(0) = \mathbf{0}$$

give $\quad y_1 = \cos 2t, \ y_2 = 2 \cos 2t \ .$

Such a solution is called a NORMAL MODE of the system $\ddot{\mathbf{y}} = B\mathbf{y}$. In Example 4.4, B is a 2×2 matrix and there are 2 normal modes. The general solution is a combination of normal modes.

Another feature of Example 4.4 is the appearance of a *symmetric* matrix which is common in mechanical problems by virtue of Newton's third law (action and reaction). An $n \times n$ symmetric matrix has the following important properties (see Murdoch, 1970).

(i) It has n *real* eigenvalues.

(ii) It has n *independent* eigenvectors.

(iii) Eigenvectors corresponding to distinct eigenvalues are orthogonal.

Problems

1. Figure 4.2 shows a mechanical system in motion at time t. The positions of the trucks are measured from the equilibrium position. Show that

$$\begin{bmatrix} \ddot{y}_1 \\ \ddot{y}_2 \end{bmatrix} = \begin{bmatrix} -(k_1+k_2)/m_1 & k_2/m_1 \\ k_2/m_2 & -(k_2+k_3)/m_2 \end{bmatrix} \begin{bmatrix} y_1 \\ y_2 \end{bmatrix} .$$

Given that $k_1 = k_2 = k_3 = m_1 = m_2 = 1$, $y_1(0) = 0$, $y_2(0) = 0$, $\dot{y}_1(0) = 0$, $\dot{y}_2(0) = 6$, find y_1 and y_2 in terms of t .

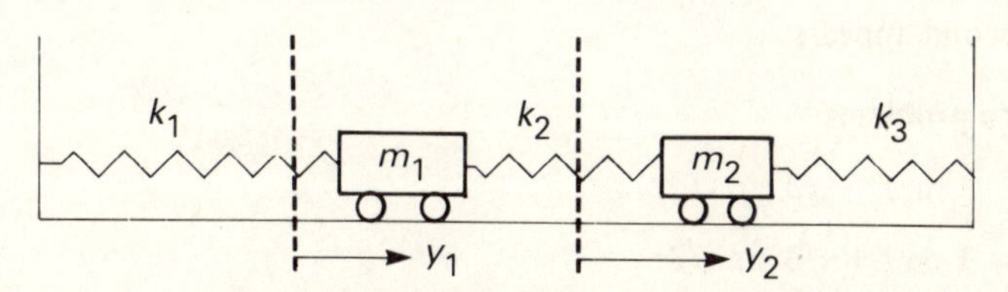

Fig. 4.2 – k_1, k_2 and k_3 are spring stiffnesses, m_1 and m_2 are masses of trucks.

2. Figure 4.3 shows three particles, each of unit mass, equally spaced on a light elastic string which is fastened rigidly by the ends and rests on a smooth horizontal table. For small displacements x_1, x_2, x_3 the equations of motion are given by

$$\ddot{\mathbf{x}} = \begin{bmatrix} 2 & -1 & 0 \\ -1 & 2 & -1 \\ 0 & -1 & 2 \end{bmatrix} \mathbf{x} .$$

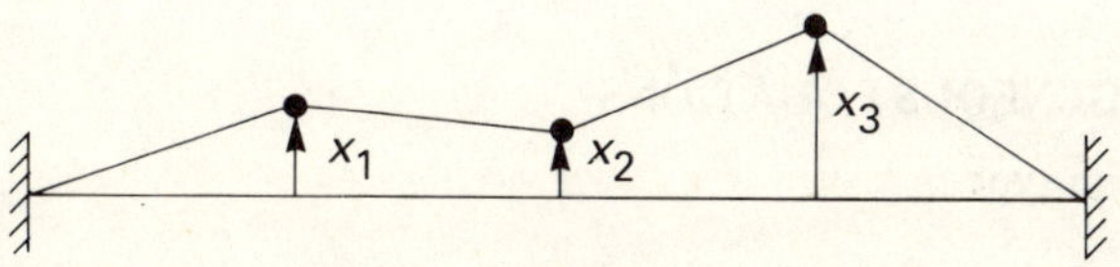

Fig. 4.3

Show that the normal modes are given by

(i)

$$\omega_1^2 = 2-\sqrt{2}, \quad \mathbf{v}_1 = \begin{bmatrix} 1 \\ 2 \\ 1 \end{bmatrix},$$

(ii)

$$\omega_2^2 = 2, \quad \mathbf{v}_2 = \begin{bmatrix} 1 \\ 0 \\ -1 \end{bmatrix},$$

(iii)

$$\omega_3^2 = 2+\sqrt{2}, \quad \mathbf{v}_3 = \begin{bmatrix} 1 \\ -2 \\ 1 \end{bmatrix},$$

and write down the general solution.

Choose an initial condition, such that only the mode of frequency $\sqrt{2}$ radians/second appears.

Answers to problems

1. $y_1 = 3 \sin t - \sqrt{3} \sin \sqrt{3}t$.

   $y_2 = 3 \sin t + \sqrt{3} \sin \sqrt{3}t$.

2. $x_1(0) = K, x_2(0) = 0, x_3(0) = -K, \dot{x}_1(0) = L$

   $\dot{x}_2(0) = 0, \dot{x}_3(0) = -L,$

for which the solution is

$$x_1 = K \cos \sqrt{2}t + \frac{L}{\sqrt{2}} \sin \sqrt{2}t$$

$$x_2 = 0$$

$$x_3 = -K \cos \sqrt{2}t - \frac{L}{\sqrt{2}} \sin \sqrt{2}t ,$$

where K and L are *small* arbitrary constants.

4.4 INHOMOGENEOUS EQUATIONS

Equations of the type

$$\dot{\mathbf{x}} = A\mathbf{x} + \mathbf{f}(t) \tag{4.2}$$

where $\mathbf{f}(t)$ is a vector whose components are functions of t, occur commonly in physical situations. Provided that the components of $\mathbf{f}(t)$ belong to the set of functions which arise in the solution of linear differential equations with constant coefficients (i.e. polynomials, exponentials, sines, cosines and their products), then equation 4.2 can be reduced to homogeneous form

$$\dot{\mathbf{x}} = A'\mathbf{x} \; . \tag{4.3}$$

The matrix A' in equation 4.3 is of the partitioned form

$$A' = \left[\begin{array}{c|c} A & B \\ \hline 0 & P \end{array}\right] ,$$

and the vector $\mathbf{x}$ in equation 4.3 is formed by adding extra components to the vector $\mathbf{x}$ of equation 4.2.

The method of solution of equation 4.2 is explained in the examples, but firstly the following theorem is useful.

Theorem 1

If the matrix A' can be partitioned as $A' = \left[\begin{array}{c|c} A & B \\ \hline 0 & P \end{array}\right]$,

where A and P are square matrices and 0 is a zero matrix, then the eigenvalues of A' can be obtained by finding the eigenvalues of the matrices A and P.

The proof of this theorem is left as an exercise for the reader, but physical considerations make it clear that the *forced* system of equation 4.2 must include terms from the unforced system $\dot{\mathbf{x}} = A\mathbf{x}$.

Example 4.5
Solve the system

$$\begin{bmatrix} \dot{x}_1 \\ \dot{x}_2 \end{bmatrix} = \begin{bmatrix} 2 & 0 \\ 1 & 1 \end{bmatrix} \begin{bmatrix} x_1 \\ x_2 \end{bmatrix} + \begin{bmatrix} 2 \\ 1 \end{bmatrix} .$$

Let $x_3 = 1$ and consider the enlarged vector $\begin{bmatrix} x_1 \\ x_2 \\ x_3 \end{bmatrix}$.

Therefore $\dot{x}_1 = 2x_1 + 2x_3$

$\dot{x}_2 = x_1 + x_2 + x_3$ from the given equations .

Also $\dot{x}_3 = 0.$

Therefore denoting the enlarged vector by x we have the homogeneous system

$$\dot{\mathbf{x}} = \left[\begin{array}{cc|c} 2 & 0 & 2 \\ 1 & 1 & 1 \\ \hline 0 & 0 & 0 \end{array}\right] \mathbf{x} = A'\mathbf{x} \ .$$

The eigenvalues of the 3 X 3 matrix A' are obtained following Theorem 4.1 from the eigenvalues of

$$\begin{bmatrix} 2 & 0 \\ 1 & 1 \end{bmatrix} \quad \text{and} \quad [0] \ ,$$

i.e. $\lambda_1 = 1, \ \lambda_2 = 2, \ \lambda_3 = 0$.

Corresponding eigenvectors of A' are

$$\mathbf{v}_1 = \begin{bmatrix} 0 \\ 1 \\ 0 \end{bmatrix}, \quad \mathbf{v}_2 = \begin{bmatrix} 1 \\ 1 \\ 0 \end{bmatrix}, \quad \mathbf{v}_3 = \begin{bmatrix} 1 \\ 0 \\ -1 \end{bmatrix} .$$

Therefore the solution of $\dot{\mathbf{x}} = A'\mathbf{x}$ is

$$\begin{bmatrix} x_1 \\ x_2 \\ x_3 \end{bmatrix} = c_1 e^t \begin{bmatrix} 0 \\ 1 \\ 0 \end{bmatrix} + c_2 e^{2t} \begin{bmatrix} 1 \\ 1 \\ 0 \end{bmatrix} + c_3 \begin{bmatrix} 1 \\ 0 \\ -1 \end{bmatrix} .$$

The solution of the given two-dimensional system can be extracted from this but, since it is two-dimensional, must only contain *two* arbitrary constants. However, we know that $x_3 = 1$, and from the above solution $x_3 = -c_3$. Therefore $c_3 = -1$ and the solution of the given system is

$$\begin{bmatrix} x_1 \\ x_2 \end{bmatrix} = c_1 e^t \begin{bmatrix} 0 \\ 1 \end{bmatrix} + c_2 e^{2t} \begin{bmatrix} 1 \\ 1 \end{bmatrix} - \begin{bmatrix} 1 \\ 0 \end{bmatrix} .$$

Example 4.6
Solve the system

$$\begin{bmatrix} \dot{x}_1 \\ \dot{x}_2 \end{bmatrix} = \begin{bmatrix} 5 & 4 \\ -1 & 0 \end{bmatrix} \begin{bmatrix} x_1 \\ x_2 \end{bmatrix} + \begin{bmatrix} 0 \\ e^{-t} \end{bmatrix}$$

Let $x_3 = \mathrm{e}^{-t}$ and note that

$$\dot{x}_3 = -\,\mathrm{e}^{-t} = -x_3 \; .$$

Let $\mathbf{x}$ represent the enlarged vector $\begin{bmatrix} x_1 \\ x_2 \\ x_3 \end{bmatrix}$.

Therefore $$\begin{bmatrix} \dot{x}_1 \\ \dot{x}_2 \\ \dot{x}_3 \end{bmatrix} = \begin{bmatrix} 5 & 4 & 0 \\ -1 & 0 & 1 \\ 0 & 0 & -1 \end{bmatrix} \begin{bmatrix} x_1 \\ x_2 \\ x_3 \end{bmatrix},$$

i.e $$\dot{\mathbf{x}} = A'\mathbf{x} \text{ where } A' = \left[\begin{array}{cc:c} 5 & 4 & 0 \\ -1 & 0 & 1 \\ \hdashline 0 & 0 & -1 \end{array}\right]$$

The eigenvalues of A' are $\lambda_1 = 4$ and $\lambda_2 = 1$, the eigenvalues of the given system

matrix $A = \begin{bmatrix} 5 & 4 \\ -1 & 0 \end{bmatrix}$, and $\lambda_3 = -1$, the eigenvalue of the matrix $[-1]$.

The eigenvectors of A' are found as follows.

For $\lambda_1 = 4$, $$(A - \lambda_1 I)\mathbf{v}_1 = \begin{bmatrix} 1 & 4 & 0 \\ -1 & -4 & 1 \\ 0 & 0 & -5 \end{bmatrix} \mathbf{v}_1 = \begin{bmatrix} 0 \\ 0 \\ 0 \end{bmatrix}.$$

Therefore $\mathbf{v}_1 = \begin{bmatrix} 4 \\ -1 \\ 0 \end{bmatrix}$ corresponds to $\lambda_1 = 4$.

Note that $\begin{bmatrix} 4 \\ -1 \end{bmatrix}$ is an eigenvector of A corresponding to $\lambda_1 = 4$.

For $\lambda_2 = 1$, $(A - \lambda_2 I)\mathbf{v}_2 = \begin{bmatrix} 4 & 4 & 0 \\ -1 & -1 & 1 \\ 0 & 0 & -2 \end{bmatrix} \mathbf{v}_2 = \begin{bmatrix} 0 \\ 0 \\ 0 \end{bmatrix}$.

Therefore $\mathbf{v}_2 = \begin{bmatrix} 1 \\ -1 \\ 0 \end{bmatrix}$ corresponds to $\lambda_2 = 1$.

Note that $\begin{bmatrix} 1 \\ -1 \end{bmatrix}$ is an eigenvector of A corresponding to $\lambda_2 = 1$.

For $\lambda_3 = -1$, $(A - \lambda_3 I)\mathbf{v}_3 = \begin{bmatrix} 6 & 4 & 0 \\ -1 & 1 & 1 \\ 0 & 0 & 0 \end{bmatrix} \mathbf{v}_3 = \begin{bmatrix} 0 \\ 0 \\ 0 \end{bmatrix}$.

Therefore $\mathbf{v}_3 = \begin{bmatrix} 2 \\ -3 \\ 5 \end{bmatrix}$ corresponds to $\lambda_3 = -1$.

Therefore the general solution of $\dot{\mathbf{x}} = A'\mathbf{x}$ is

$$\begin{bmatrix} x_1 \\ x_2 \\ x_3 \end{bmatrix} = c_1 e^{4t} \begin{bmatrix} 4 \\ -1 \\ 0 \end{bmatrix} + c_2 e^{t} \begin{bmatrix} 1 \\ -1 \\ 0 \end{bmatrix} + c_3 e^{-t} \begin{bmatrix} 2 \\ -3 \\ 5 \end{bmatrix} .$$

However, this cannot be the general solution of the original two-dimensional system, since this requires *only two* arbitrary constants.

The above solution gives $x_3 = 5\,c_3e^{-t}$, but we chose $x_3 = e^{-t}$. Therefore we must have $c_3 = \frac{1}{5}$. It follows that

$$\begin{bmatrix} x_1 \\ x_2 \end{bmatrix} = c_1e^{4t}\begin{bmatrix} 4 \\ -1 \end{bmatrix} + c_2e^{t}\begin{bmatrix} 1 \\ -1 \end{bmatrix} + e^{-t}\begin{bmatrix} 2/5 \\ -3/5 \end{bmatrix}$$

is the general solution of the given system.

Example 4.7
Solve the system

$$\begin{bmatrix} \dot{x}_1 \\ \dot{x}_2 \end{bmatrix} = \begin{bmatrix} 0 & 1 \\ 2 & 1 \end{bmatrix}\begin{bmatrix} x_1 \\ x_2 \end{bmatrix} + \begin{bmatrix} \sin t \\ 2\cos t \end{bmatrix}.$$

Let $x_3 = \sin t$ and $x_4 = \cos t$. Note that $\dot{x}_3 = x_4, \dot{x}_4 = -x_3$.

Letting $\mathbf{x} = \begin{bmatrix} x_1 \\ x_2 \\ x_3 \\ x_4 \end{bmatrix}$, we have

$$\begin{bmatrix} \dot{x}_1 \\ \dot{x}_2 \\ \dot{x}_3 \\ \dot{x}_4 \end{bmatrix} = \begin{bmatrix} 0 & 1 & 1 & 0 \\ 2 & 1 & 0 & 2 \\ 0 & 0 & 0 & 1 \\ 0 & 0 & -1 & 0 \end{bmatrix}\begin{bmatrix} x_1 \\ x_2 \\ x_3 \\ x_4 \end{bmatrix},$$

i.e. $\dot{\mathbf{x}} = A'\mathbf{x}$ where $A' = \left[\begin{array}{cc|cc} 0 & 1 & 1 & 0 \\ 2 & 1 & 0 & 2 \\ \hline 0 & 0 & 0 & 1 \\ 0 & 0 & -1 & 0 \end{array}\right]$.

The eigenvalues of A' are

$$\lambda_1 = -1,\ \lambda_2 = 2,\ \text{the eigenvalues of } \begin{bmatrix} 0 & 1 \\ 2 & 1 \end{bmatrix},$$

and $\lambda_3 = j,\ \lambda_4 = -j$, the eigenvalues of $\begin{bmatrix} 0 & 1 \\ -1 & 0 \end{bmatrix}$.

The eigenvectors of A' are found as follows.

For $\lambda_1 = -1$, $$\begin{bmatrix} 1 & 1 & 1 & 0 \\ 2 & 2 & 0 & 2 \\ 0 & 0 & 1 & 1 \\ 0 & 0 & -1 & 1 \end{bmatrix} \mathbf{v}_1 = \begin{bmatrix} 0 \\ 0 \\ 0 \\ 0 \end{bmatrix}.$$

Therefore $\mathbf{v}_1 = \begin{bmatrix} 1 \\ -1 \\ 0 \\ 0 \end{bmatrix}$. Note that $\begin{bmatrix} 1 \\ -1 \end{bmatrix}$ is eigenvector of $\begin{bmatrix} 0 & 1 \\ 2 & 1 \end{bmatrix}$.

For $\lambda_2 = 2$, $$\begin{bmatrix} -2 & 1 & 1 & 0 \\ 2 & -1 & 0 & 2 \\ 0 & 0 & -2 & 1 \\ 0 & 0 & -1 & -2 \end{bmatrix} \mathbf{v}_2 = \begin{bmatrix} 0 \\ 0 \\ 0 \\ 0 \end{bmatrix}.$$

Therefore $\mathbf{v}_2 = \begin{bmatrix} 1 \\ 2 \\ 0 \\ 0 \end{bmatrix}$. Note that $\begin{bmatrix} 1 \\ 2 \end{bmatrix}$ is eigenvector of $\begin{bmatrix} 0 & 1 \\ 2 & 1 \end{bmatrix}$.

For $\lambda_3 = j$, $$\begin{bmatrix} -j & 1 & 1 & 0 \\ 2 & 1-j & 0 & 2 \\ 0 & 0 & -j & 1 \\ 0 & 0 & -1 & -j \end{bmatrix} \mathbf{v}_3 = \begin{bmatrix} 0 \\ 0 \\ 0 \\ 0 \end{bmatrix}.$$

Therefore $\mathbf{v}_3 = \begin{bmatrix} 1 \\ 0 \\ j \\ -1 \end{bmatrix}$ is eigenvector corresponding to $\lambda_3 = j$.

Therefore $\mathbf{x} = e^{jt} \begin{bmatrix} 1 \\ 0 \\ j \\ -1 \end{bmatrix} = (\cos t + j \sin t) \begin{bmatrix} 1 \\ 0 \\ j \\ -1 \end{bmatrix}$

is a solution of $\dot{\mathbf{x}} = A'\mathbf{x}$.

Therefore $\mathbf{x} = \begin{bmatrix} \cos t \\ 0 \\ -\sin t \\ -\cos t \end{bmatrix}$ and $\mathbf{x} = \begin{bmatrix} \sin t \\ 0 \\ \cos t \\ -\sin t \end{bmatrix}$

are independent solutions of $\dot{\mathbf{x}} = A'\mathbf{x}$ when $\lambda = \pm j$.
Therefore the general solution of $\dot{\mathbf{x}} = A'\mathbf{x}$ is

$$\begin{bmatrix} x_1 \\ x_2 \\ x_3 \\ x_4 \end{bmatrix} = c_1 e^{-t} \begin{bmatrix} 1 \\ -1 \\ 0 \\ 0 \end{bmatrix} + c_2 e^{2t} \begin{bmatrix} 1 \\ 2 \\ 0 \\ 0 \end{bmatrix} + c_3 \begin{bmatrix} \cos t \\ 0 \\ -\sin t \\ -\cos t \end{bmatrix} + c_4 \begin{bmatrix} \sin t \\ 0 \\ \cos t \\ -\sin t \end{bmatrix} .$$

This gives $x_3 = -c_3 \sin t + c_4 \cos t$,

$x_4 = -c_3 \cos t - c_4 \sin t$;

but $x_3 = \sin t$, i.e. $c_3 = -1$,

$x_4 = \cos t$, $c_4 = 0$.

Therefore the general solution of the given system is

$$\begin{bmatrix} x_1 \\ x_2 \end{bmatrix} = c_1 e^{-t} \begin{bmatrix} 1 \\ -1 \end{bmatrix} + c_2 e^{2t} \begin{bmatrix} 1 \\ 2 \end{bmatrix} + \begin{bmatrix} -\cos t \\ 0 \end{bmatrix} .$$

The special case of equation 4.2 when $\mathbf{f}(t) = \mathbf{b}$, a constant vector, can readily be solved using an alternative method similar to the techniques used for the solution of simple linear differential equations with constant coefficients.

Given the system

$$\dot{\mathbf{x}} = A\mathbf{x} + \mathbf{b} \tag{4.4}$$

we begin by solving the corresponding homogeneous equation

$$\dot{\mathbf{x}} = A\mathbf{x} \ . \tag{4.5}$$

Then the solution of equation 4.4 is given by adding the general solution of equation 4.5 to *any* particular solution of equation 4.4. However, $\mathbf{x} = -A^{-1}\mathbf{b}$ is a solution of equation 4.4 provided A is a non-singular matrix. Therefore $\mathbf{x} = \mathbf{x}_H - A^{-1}\mathbf{b}$ is the solution of equation 4.4 (where $\mathbf{x} = \mathbf{x}_H$ is the general solution of the homogeneous system).

Example 4.8

Repeat Example 4.5 using the alternative method.

$$\dot{\mathbf{x}} = \begin{bmatrix} 2 & 0 \\ 1 & 1 \end{bmatrix} \mathbf{x} + \begin{bmatrix} 2 \\ 1 \end{bmatrix} \tag{4.6}$$

The eigenvalues and eigenvectors of the matrix $A = \begin{bmatrix} 2 & 0 \\ 1 & 1 \end{bmatrix}$ are

$$\lambda_1 = 1, \ \mathbf{v}_1 = \begin{bmatrix} 0 \\ 1 \end{bmatrix}, \ \lambda_2 = 2, \mathbf{v}_2 = \begin{bmatrix} 1 \\ 1 \end{bmatrix}.$$

Therefore the general solution of the homogeneous system $\dot{\mathbf{x}} = A\mathbf{x}$ is

$$\mathbf{x}_H = c_1 e^t \begin{bmatrix} 0 \\ 1 \end{bmatrix} + c_2 e^{2t} \begin{bmatrix} 1 \\ 1 \end{bmatrix}.$$

A particular solution of equation 4.6 is given by

$$\mathbf{x} = -A^{-1} \begin{bmatrix} 2 \\ 1 \end{bmatrix}$$

$$= -\tfrac{1}{2} \begin{bmatrix} 1 & 0 \\ -1 & 2 \end{bmatrix} \begin{bmatrix} 2 \\ 1 \end{bmatrix}$$

$$= - \begin{bmatrix} 1 \\ 0 \end{bmatrix}.$$

Therefore the general solution of equation 4.6 is

$$\mathbf{x} = c_1 e^t \begin{bmatrix} 0 \\ 1 \end{bmatrix} + c_2 e^{2t} \begin{bmatrix} 1 \\ 1 \end{bmatrix} - \begin{bmatrix} 1 \\ 0 \end{bmatrix} .$$

It should be noted that equation 4.6 has an equilibrium point at (−1,0) in the $x_1 x_2$ plane, since

$$\dot{\mathbf{x}} = \begin{bmatrix} 2 & 0 \\ 1 & 1 \end{bmatrix} \begin{bmatrix} -1 \\ 0 \end{bmatrix} + \begin{bmatrix} 2 \\ 1 \end{bmatrix} = \begin{bmatrix} 0 \\ 0 \end{bmatrix}$$

at this point.

The nature of the equilibrium point is determined by considering the eigenvalues of the matrix

$$A = \begin{bmatrix} 2 & 0 \\ 1 & 1 \end{bmatrix}$$

as in Section 3.4. For equation 4.6 the point (−1,0) is an *unstable node.*

In general if A is a non-singular 2 × 2 matrix the equation

$$\dot{\mathbf{x}} = A\mathbf{x} + \mathbf{b} \tag{4.7}$$

has an equilibrium point given by the vector

$$\mathbf{x}_E = -A^{-1}\mathbf{b} \ . \tag{4.8}$$

The substitution $\mathbf{X} = \mathbf{x} - \mathbf{x}_E$ in equation 4.7 gives the homogeneous system

$$\dot{\mathbf{X}} = A\mathbf{X} \tag{4.9}$$

which has its equilibrium point at the origin $\mathbf{X} = \mathbf{0}$. Clearly the nature of the equilibrium points $\mathbf{X} = \mathbf{0}$ for equation 4.9 and $\mathbf{x} = \mathbf{x}_E$ for equation 4.7 is identical. Thus the results of Section 3.4 for the two-dimensional homogeneous system can be applied to the system given by equation 4.7.

Problems

1. Solve the following systems.

(i)

$$\dot{\mathbf{x}} = \begin{bmatrix} 0.6 & 0.8 \\ 0.8 & -0.6 \end{bmatrix} \mathbf{x} + \begin{bmatrix} -1 \\ 2 \end{bmatrix}$$

(ii)

$$\dot{\mathbf{x}} = \begin{bmatrix} 1 & 0 \\ 1 & 1 \end{bmatrix} \mathbf{x} + \begin{bmatrix} 2 \\ 3 \end{bmatrix}$$

(iii)

$$\dot{\mathbf{x}} = \begin{bmatrix} -9 & 2 & 6 \\ 5 & 0 & -3 \\ -16 & 4 & 11 \end{bmatrix} \mathbf{x} + \begin{bmatrix} 1 \\ 2 \\ 3 \end{bmatrix}$$

(iv)

$$\mathbf{x} = \begin{bmatrix} -3 & 1 & 0 \\ 0 & -3 & 1 \\ 4 & -8 & 2 \end{bmatrix} \mathbf{x} + \begin{bmatrix} 1 \\ 1 \\ 2 \end{bmatrix}$$

2. Solve the system

$$\begin{aligned} \dot{x}_1 &= x_2 + t + 1 \\ \dot{x}_2 &= 2x_1 + x_2 + 3 \ . \end{aligned}$$

3. Solve the initial value problem

$$\begin{bmatrix} \dot{x}_1 \\ \dot{x}_2 \end{bmatrix} = \begin{bmatrix} 0 & 1 \\ 2 & 1 \end{bmatrix} \begin{bmatrix} x_1 \\ x_2 \end{bmatrix} + \begin{bmatrix} e^t \\ \cos t \end{bmatrix}, \quad \begin{bmatrix} x_1(0) \\ x_2(0) \end{bmatrix} = \begin{bmatrix} -1 \\ 1.7 \end{bmatrix} .$$

4. Find the position and nature of the equilibrium points for the following systems.

(i)

$$\dot{\mathbf{x}} = \begin{bmatrix} 1 & 1 \\ 4 & 1 \end{bmatrix} \mathbf{x} + \begin{bmatrix} 1 \\ 1 \end{bmatrix}$$

(ii)

$$\dot{\mathbf{x}} = \begin{bmatrix} -2 & -5 \\ 1 & 0 \end{bmatrix} \mathbf{x} + \begin{bmatrix} 3 \\ -2 \end{bmatrix}$$

(iii)

$$\dot{\mathbf{x}} = \begin{bmatrix} -1 & 2 \\ 7 & 1 \end{bmatrix} \mathbf{x} + \begin{bmatrix} 1 \\ 7 \end{bmatrix}$$

(iv)

$$\dot{\mathbf{x}} = \begin{bmatrix} -2 & -2 \\ 3 & 2 \end{bmatrix} \mathbf{x} - \begin{bmatrix} 0 \\ 3 \end{bmatrix}$$

5. Solve the system

$$\begin{aligned} \ddot{x}_1 &= -10x_1 + 6x_2 + e^{-t} \\ \ddot{x}_2 &= 15x_1 - 19x_2 \ . \end{aligned}$$

Answers to problems

1. (i)
$$\mathbf{x} = c_1 e^t \begin{bmatrix} 2 \\ 1 \end{bmatrix} + c_2 e^{-t} \begin{bmatrix} -1 \\ 2 \end{bmatrix} - \begin{bmatrix} 1 \\ -2 \end{bmatrix} .$$

(ii)
$$\mathbf{x} = c_1 e^t \begin{bmatrix} 0 \\ 1 \end{bmatrix} + c_2 e^t \begin{bmatrix} 1 \\ t \end{bmatrix} - \begin{bmatrix} 2 \\ 1 \end{bmatrix} .$$

(iii)
$$\mathbf{x} = c_1 e^t \begin{bmatrix} 1 \\ -1 \\ 2 \end{bmatrix} + c_2 e^{-t} \begin{bmatrix} 2 \\ -1 \\ 3 \end{bmatrix} + c_3 e^{2t} \begin{bmatrix} -2 \\ 1 \\ -4 \end{bmatrix} - \begin{bmatrix} 1 \\ 2 \\ 1 \end{bmatrix} .$$

(iv)
$$\mathbf{x} = c_1 e^{-t} \begin{bmatrix} 1 \\ 2 \\ 4 \end{bmatrix} + c_2 e^{-t} \begin{bmatrix} 1-t \\ 1-2t \\ -4t \end{bmatrix} + c_3 e^{-2t} \begin{bmatrix} 1 \\ 1 \\ 1 \end{bmatrix} + \begin{bmatrix} 1 \\ 2 \\ 5 \end{bmatrix} .$$

2. $x_1 = c_1 e^{-t} + c_2 e^{2t} - \frac{7}{4} + \frac{t}{2}$.
$x_2 = -c_1 e^{-t} + 2c_2 e^{2t} - \frac{1}{2} - t$.

3. $x_1 = -\frac{7}{5} e^{-t} + \frac{7}{10} e^{2t} - \frac{1}{10}(\sin t + 3 \cos t)$.
$x_2 = \frac{7}{5} e^{-t} + \frac{7}{5} e^{2t} - \frac{1}{10}(10e^t - 3 \sin t + \cos t)$.

4. (i) $(0, -1)$, saddle .
(ii) $(2, -\frac{1}{5})$, stable focus .
(iii) $(-\frac{13}{15}, -\frac{14}{15})$, saddle .
(iv) $(3, -3)$, centre .

5. $x_1 = c_1 \cos 2t + c_2 \sin 2t + 2c_3 \cos 5t + 2c_4 \sin 5t + \frac{2}{13} e^{-t}$.
$x_2 = c_1 \cos 2t + c_2 \sin 2t - 5c_3 \cos 5t - 5c_4 \sin 5t + \frac{3}{26} e^{-t}$.

Part II

Nonlinear differential equations – graphical methods

5

Features of nonlinear differential equations

5.1 INTRODUCTION

We saw in Part I that there is an organised and well-defined theory available for the solution of linear differential equations with constant coefficients. There are of course other methods of solution, but whichever method is used the solution can be obtained by applying systematically a certain procedure. There is no such organised theory for the analytic solution of nonlinear differential equations. Methods which are useful in certain cases are not applicable in other situations. Some methods may not provide a solution at all, but will merely indicate certain properties of the solution.

In this book we shall consider almost exclusively either the second order autonomous differential equation

$$\ddot{x} + f(x, \dot{x}) = 0 \tag{5.1}$$

or the second order system

$$\begin{aligned} \dot{x} &= P(x, y) \\ \dot{y} &= Q(x, y) \end{aligned} \tag{5.2}$$

where f, P and Q are nonlinear functions.

Studies of continuous systems in engineering and science frequently lead to mathematical models in the form of one of the categories (5.1) or (5.2). In many cases analytical solution is impossible, and the use of a computer (either analog or digital) is necessary. However, it is sometimes possible to establish some properties of the solution using analytical techniques before going to the computer. These techniques do not provide a solution but can be useful in two ways.

(i) They provide an insight into the problem and may indicate restrictions on certain parameters which will reduce the number of computer runs required to complete the solution.

(ii) They provide some check for the final computer solution.

In this chapter we consider some of the properties of nonlinear equations which distinguish them from linear differential equations.

5.2 SOME PROPERTIES OF NONLINEAR DIFFERENTIAL EQUATIONS

(a) <u>The frequency of unforced oscillations may depend on the initial conditions.</u>

Consider the nonlinear equation

$$\ddot{x} + \sin x = 0 \quad . \tag{5.3}$$

When x is 'small' we have $\sin x \approx x$ and the equation is approximately

$$\ddot{x} + x = 0 \quad . \tag{5.4}$$

Equation 5.4 is linear and for the initial conditions $x(0) = x_0$, $\dot{x}(0) = 0$ its solution is $x = x_0 \cos t$, i.e. the frequency of oscillations is 1 for *all* x_0. We shall show later that the solution of equation 5.3 for the same initial conditions is approximately $x = x_0 \cos (1 - x_0^2/16)t$, i.e. the frequency depends on x_0 (see Fig. 5.1).

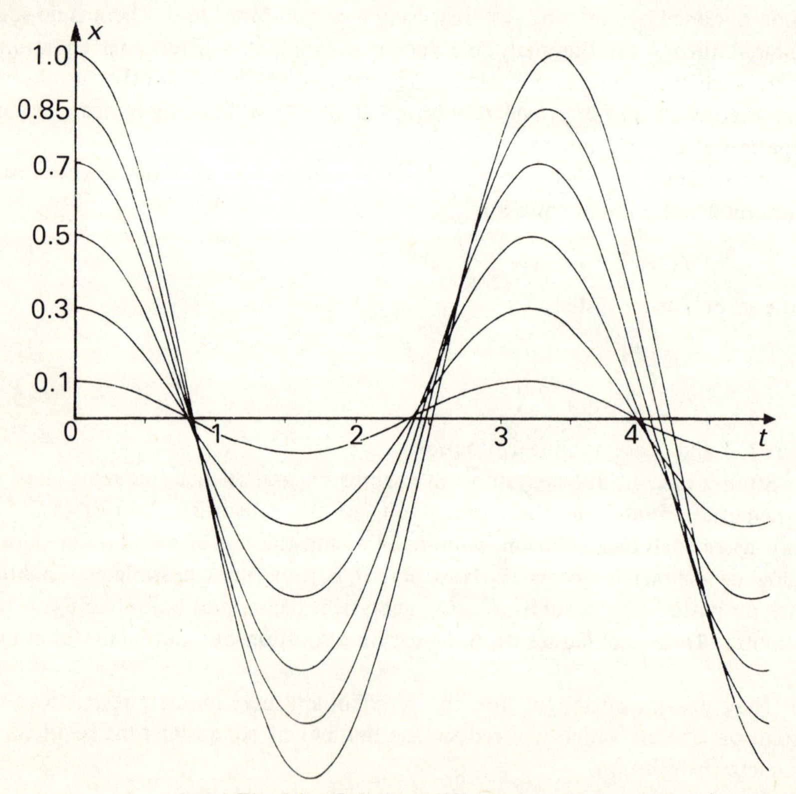

Fig. 5.1 – Solutions of $\ddot{x} + \sin x = 0$.

(b) Stability may depend on initial conditions.

Consider the nonlinear equation

$$\ddot{x} + 0.1(1-x^2)\dot{x} + x = 0 \tag{5.5}$$

and a possible linear approximation for small x

$$\ddot{x} + 0.1\dot{x} + x = 0 \quad . \tag{5.6}$$

For the initial conditions $x(0) = x_0$, $\dot{x}(0) = 0$ the two equations do have similar solutions when x_0 is small, but there are surprising differences for larger x_0. The solution of equation 5.6 is

$$x = e^{-0.05t}(x_0 \cos 0.9975t + \frac{0.05}{0.9975} x_0 \sin 0.9975t)$$

for all x_0 and we note that $x \to 0$ as $t \to \infty$.

It will be shown later that the solution of equation 5.5 is stable only if $x_0 < 2$ (approximately), i.e.

$$x \to 0 \text{ as } t \to \infty \text{ for } x_0 < 2,$$

$$x \to \infty \text{ as } t \to \infty \text{ for } x_0 > 2.$$

Typical solutions of equation 5.5 are shown in Fig. 5.2.

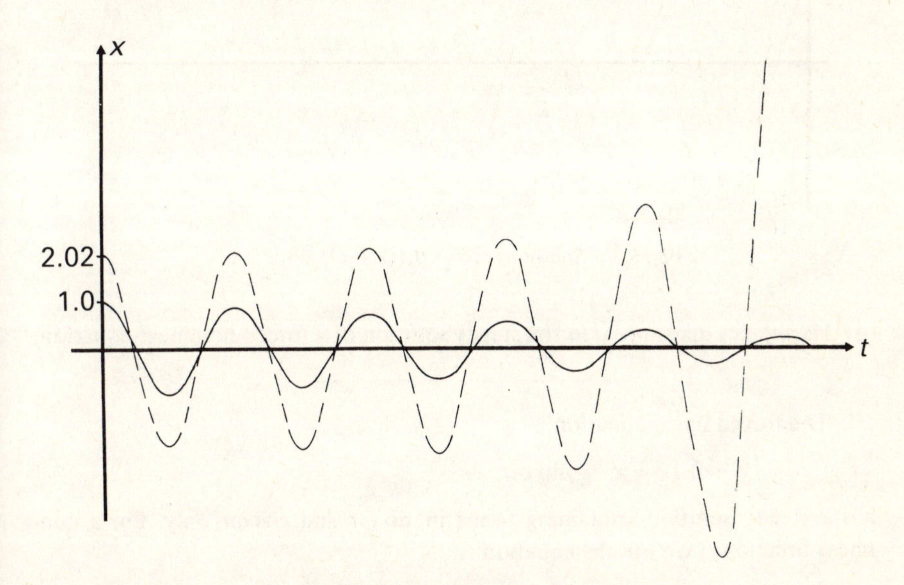

Fig. 5.2 – Solutions of $\ddot{x} + 0.1(1 - x^2)\dot{x} + x = 0$.

(c) The amplitude of periodic oscillations may be independent of initial conditions.

For the periodic solution of an unforced linear equation the amplitude depends on the initial conditions. For example the initial value problem

$$\ddot{x} + x = 0, \; x(0) = x_0, \; \dot{x}(0) = 0$$

has solution $x = x_0 \cos t$. However, the nonlinear initial value problem

$$\ddot{x} - 0.1(1 - x^2)\dot{x} + x = 0, \;\; x(0) = x_0, \dot{x}(0) = 0$$

has solution $x = 2 \cos t$, approximately, after transient terms have become negligible. Figure 5.3 shows solutions for various values of x_0.

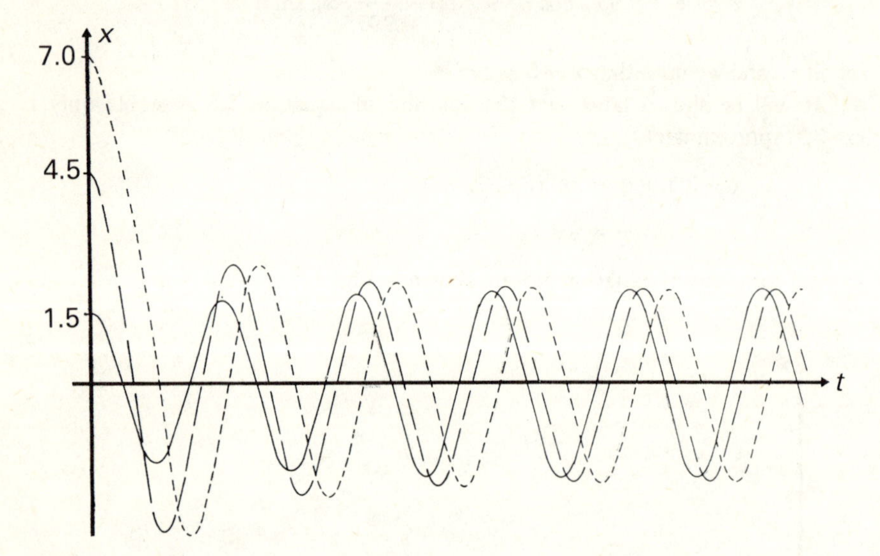

Fig. 5.3 – Solutions of $\ddot{x} - 0.1(1 - x^2)\,\dot{x} + x = 0$.

(d) Harmonics may appear in the steady solution of a forced nonlinear equation.

The forced linear equation,

$$\ddot{x} + 0.1\dot{x} + x = \sin \omega t,$$

has a steady solution containing terms in $\sin \omega t$ and $\cos \omega t$ only. For a non-linear function $f(x, \dot{x})$ in the equation

$$\ddot{x} + f(x, \dot{x}) = \sin \omega t,$$

the steady solution may contain terms in

$$\sin \omega t, \cos \omega t, \sin 2\omega t, \cos 2\omega t, \sin 3\omega t, \cos 3\omega t, \text{etc.},$$

and sometimes terms in

$$\sin \frac{\omega t}{2}, \cos \frac{\omega t}{2}, \sin \frac{\omega t}{3}, \cos \frac{\omega t}{3}, \text{etc.}$$

The examples in this section have indicated that the presence of a nonlinear term in a differential equation can lead to features in the solution very different from what we have come to expect while dealing with linear equations. It is clear that great care must be taken when making linear approximations to nonlinear differential equations. Chapter 8 deals with linearisation in more detail.

Problem

A mass, m, is suspended on a spring of stiffness, k. Assuming that Hooke's law applies, i.e. restoring force is proportional to extension x, the equation of motion is

$$m\ddot{x} = -(Kx + f(\dot{x})),$$

where $f(\dot{x})$ is a nonlinear function of $\dot{x}$ representing friction due to motion through air. Given initial conditions $x(0) = x_0, \dot{x}(0) = 0$, what would you expect the solution to be as $t \to \infty$ for the cases (i) $f(\dot{x}) = 0$, (ii) $f(\dot{x}) = 0.0001\dot{x}^3$? Is case (i) a satisfactory linear approximation to case (ii)?

6

Graphical methods of solution

6.1 INTRODUCTION

This chapter deals with two methods which require the construction of fairly accurate graphs. The method of isoclines applies to first order differential equations but is easily extended to give a solution for the second order equation $\ddot{x} + F(x, \dot{x}) = 0$ as a graph in the $x\,\dot{x}$ plane.

Liénard's method is applicable only to the special type $\ddot{x} + f(\dot{x}) + x = 0$.

6.2 THE METHOD OF ISOCLINES

This method is used to obtain graphical solutions of the first order differential equation

$$\frac{dy}{dx} = f(x,y) \quad . \tag{6.1}$$

The family of curves $f(x, y) = \lambda$ is called the set of ISOCLINES for equation 6.1. The family of solution curves for equation 6.1 (alternatively called INTEGRAL CURVES) intersect the isoclines at points where $dy/dx = \lambda$. The method of isoclines involves two stages:

(i) an accurate drawing of the isoclines,
(ii) construction of the required solution curves such that as each curve crosses an isocline the gradient is taken as the value of λ corresponding to that isocline.

The method is demonstrated for a simple linear equation in Example 6.1.

The method of isoclines can be extended to second order differential equations of the type

$$\ddot{x} + F(x, \dot{x}) = 0 \quad . \tag{6.2}$$

The substitutions $y = \dot{x}$,

$$y\frac{dy}{dx} = \frac{dx}{dt}\cdot\frac{dy}{dx} = \frac{dy}{dt} = \ddot{x} \ ,$$

reduce equation 6.2 to

$$\frac{dy}{dx} = -\frac{F(x, y)}{y}$$

which is of the form of equation 6.1. Solution curves can now be drawn in the $x\dot{x}$ plane. The $x\dot{x}$ plane is called the PHASE-PLANE, and Chapter 7 considers solutions in the phase-plane in more detail. Example 6.2 applies the isocline method to the undamped simple pendulum equation $\ddot{x} + \sin x = 0$. A graphical method for obtaining $x \sim t$ solution curves from phase-plane solutions is given at the end of Chapter 7.

Example 6.1

Use the method of isoclines to solve the differential equation

$$\frac{dy}{dx} = y, \; y(0) = 0.5 \quad .$$

The isoclines are $y = \lambda$, which are a family of straight lines parallel to the x-axis. The inital point (0,0.5) lies on the isocline $y = \frac{1}{2}$, i.e. $dy/dx = \frac{1}{2}$ at this point. y increases and the curve is drawn with the slope, dy/dx, being adjusted to correspond to each isocline. See Fig. 6.1 and compare with the analytical solution, $y = 0.5e^x$.

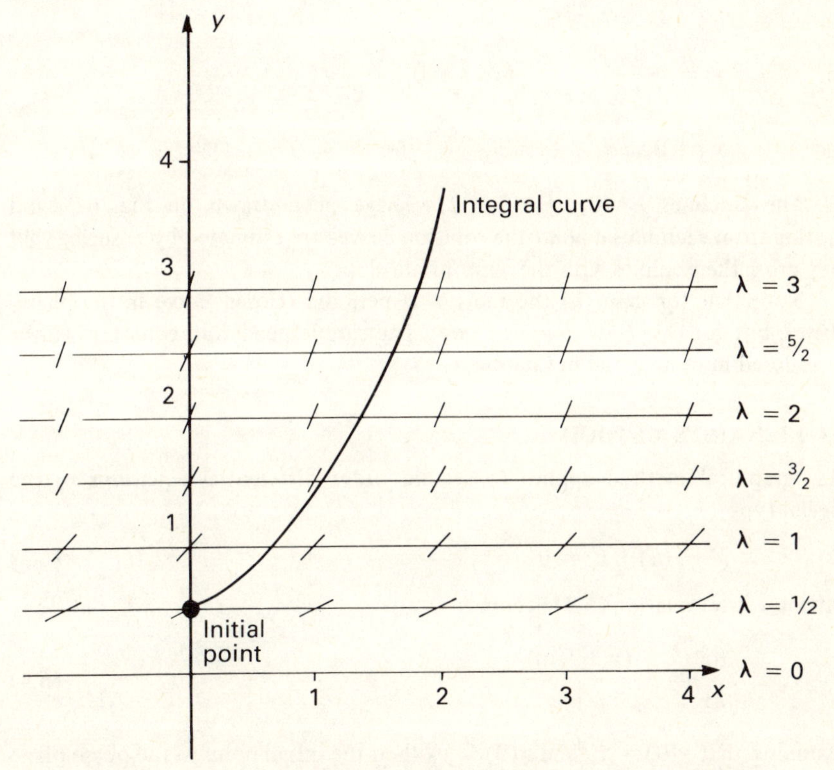

Fig. 6.1 – Solutions of $dy/dx = y$, $y(0) = 0.5$, using method of isoclines.

Example 6.2

Use the method of isoclines to draw the solution curves in the phase-plane for the equation

$$\ddot{x} + \sin x = 0$$

for the two cases

$$\text{(i)}\ x(0) = -\pi\ ,\quad \dot{x}(0) = 0.5\ ,$$

$$\text{(ii)}\ x(0) = -\frac{\pi}{2}\ ,\quad \dot{x}(0) = 0\ .$$

Putting $y = \dot{x}$ gives $\dfrac{\mathrm{d}y}{\mathrm{d}x} = -\dfrac{\sin x}{y}$.

The isoclines are the family of curves

$$-\frac{\sin x}{y} = \lambda\ ,$$

i.e.
$$y = -\frac{\sin x}{\lambda} \qquad \text{for } \lambda \neq 0,$$

and
$$x = 0,\ \pm\pi,\ \pm 2\pi, \ldots \qquad \text{for } \lambda = 0\ .$$

The isoclines $\lambda = 0, \pm\frac{1}{2}, \pm 1, \pm 2, \infty$ have been drawn on Fig. 6.2, and starting from each initial point the solution curves are estimated by ensuring that they cross the isoclines with the appropriate slope.

Note that for case (ii) the motion is periodic (closed curve in the phase-plane), but for case (i) $x \to \infty$ as $t \to \infty$. The simple pendulum equation will be considered in more detail in Chapter 7.

6.3 LIÉNARD'S METHOD

This graphical method applies to second order differential equations of the special type

$$\ddot{x} + f(\dot{x}) + x = 0\ . \tag{6.3}$$

Putting $y = \dot{x}$ equation (6.3) becomes

$$\frac{\mathrm{d}y}{\mathrm{d}x} = -\frac{(x + f(y))}{y}\ . \tag{6.4}$$

Assuming that $x(0) = x_0$ and $\dot{x}(0) = y_0$ then the initial point in the phase-plane is (x_0, y_0).

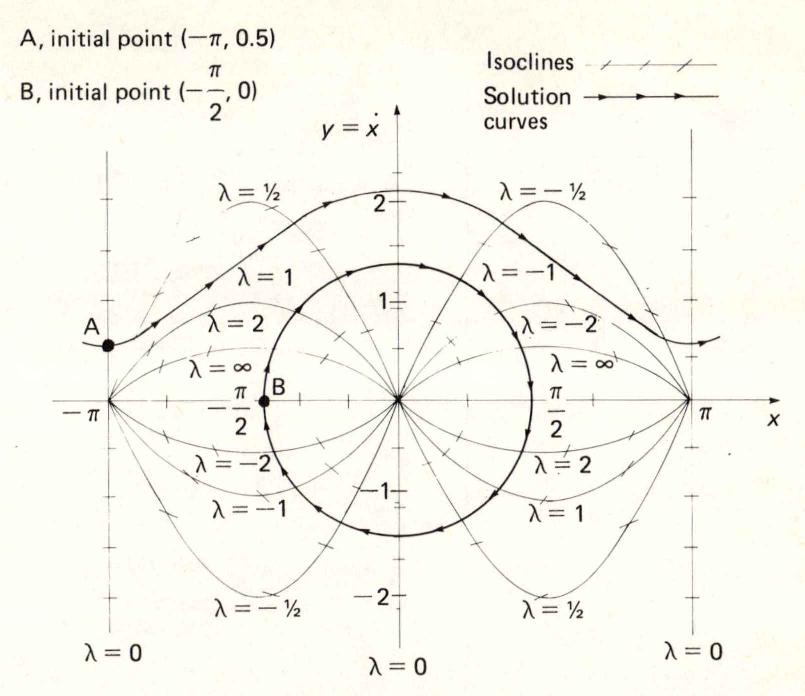

Fig. 6.2 – Phase-plane solution of $\ddot{x} + \sin x = 0$.

Liénard's method starts by drawing the curve $x + f(y) = 0$ in the phase-plane (see Fig. 6.3). A line parallel to the x-axis is drawn from the initial point $P_0(x_0, y_0)$, to intersect $x + f(y) = 0$ at Q. QR, perpendicular to the x-axis is drawn, cutting the x-axis at R. It is easily shown that the gradient of the line RP_0 is

$$\frac{y_0}{x_0 + f(y_0)} = -\frac{1}{\left(\frac{dy}{dx}\right)_0} .$$

Therefore, the integral curve through P_0 is perpendicular to RP_0. A small segment P_0P_1 is drawn perpendicular to RP_0 and the process is repeated starting from P_1, (see Fig. 6.3). The method is demonstrated in Example 6.3.

Example 6.3

Use Liénard's method to draw the solution curve in the phase-plane for the equation

$$\ddot{x} + \dot{x}\,|\dot{x}| + x = 0, \quad x(0) = 1, \quad \dot{x}(0) = 0.5 .$$

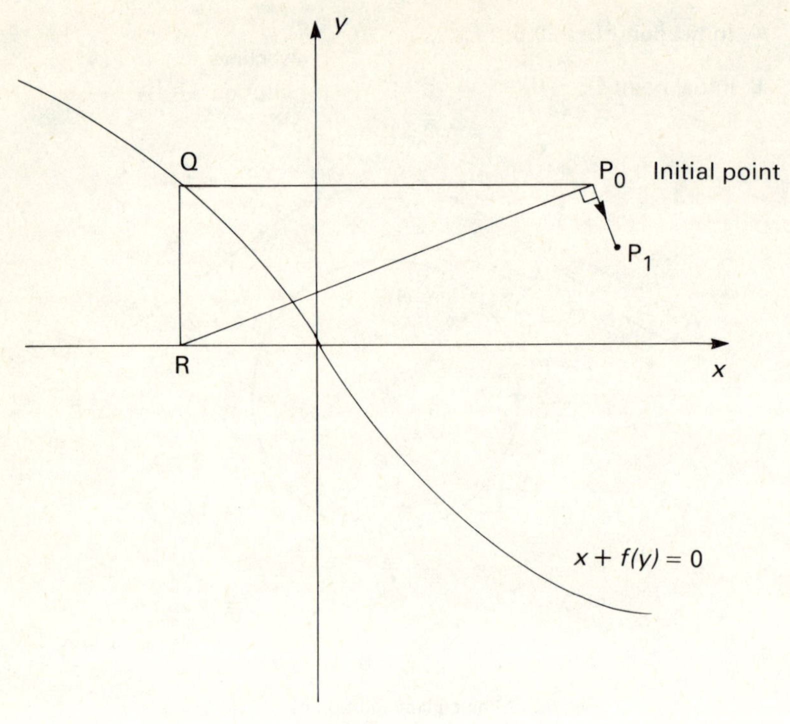

Fig. 6.3 – Liénard's method.

Putting $y = \dot{x}$ gives

$$\frac{dy}{dx} = -\frac{x + y\,|\,y\,|}{y} \;.$$

Therefore the curve $x + y\,|\,y\,| = 0$ is drawn in the phase-plane,

i.e

$$y^2 = \begin{cases} x & \text{for} \quad y < 0 \\ -x & \text{for} \quad y > 0 \;. \end{cases}$$

Figure 6.4 shows five steps of Liénard's method. Clearly the accuracy of the graphical solution can be improved by taking smaller increments.

Problems

1. Sketch the isoclines and portions of the solution curves for the differential equation

$$\frac{dy}{dx} = y + 0.1\,x^2 \;.$$

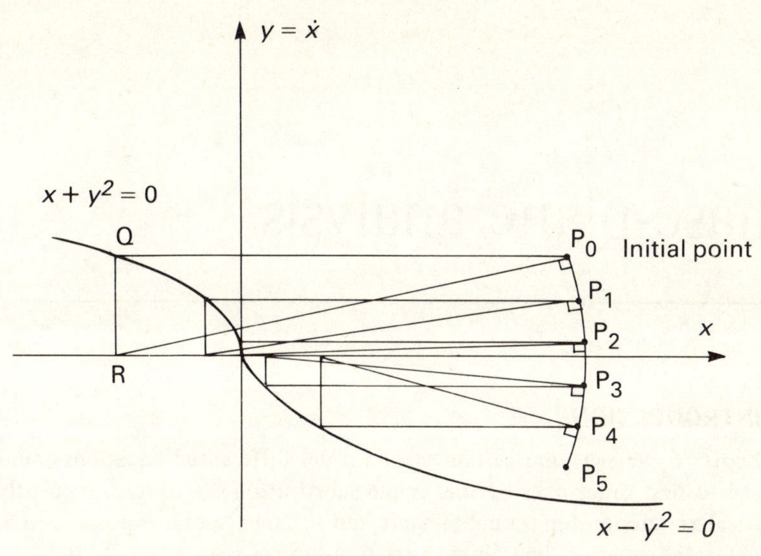

Fig. 6.4 – Solution of $\ddot{x} + \dot{x}\,|\dot{x}| + x = 0$ using Liénard's method.

2. Use the method of isoclines to sketch the solution curves for the equation

$$\frac{dy}{dx} = x^2 + y^2 \;.$$

Verify that if $y(0) = 0$, then $y(1) \approx 0.35$.

3. Use isoclines to draw the solution curve in the phase-plane for the initial value problem

$$\ddot{x} + 2\dot{x} + x = 0, \; x(0) = 2, \; \dot{x}(0) = 0 \;.$$

Compare with the analytical solution.

4. Sketch the phase-plane diagram of the motion given by the differential equation

$$\ddot{x} + \dot{x}^2 - x\,\dot{x} = 0 \text{ with initial conditions } x(0) = 0, \; \dot{x}(0) = 1.$$

5. Use Liénard's method to obtain the solution to the initial value problem of Problem 3.

6. Rayleigh's equation is

$$\ddot{x} - \epsilon(\dot{x} - \tfrac{1}{3}\dot{x}^3) + x = 0 \;.$$

For the case $\epsilon = 1$, use Liénard's method to draw the solution curve in the phase-plane which passes through the point $x = 1$, $\dot{x} = 0$. Comment on the steady motion.

7

Phase-plane analysis

7.1 INTRODUCTION

In Chapter 6 we saw that certain second order differential equations could be reduced to first order ones by the simple substitution $y = dx/dt$. Frequently an analytical xt relationship is unobtainable, but in some cases it is possible to solve the first order equations and obtain a relationship between x and dx/dt.

By this method it may be possible to obtain valuable information about the second order equation's solution, even when the full solution is not known.

7.2 THE PHASE-PLANE

The $x\dot{x}$ plane is known as the PHASE-PLANE and the set of solutions in this plane (for different initial conditions) is called the PHASE-PLANE DIAGRAM.

The direction of motion of the point $(x, \dot{x})$ along a solution curve in the phase-plane is determined simply from the fact that x increases in the upper half-plane ($\dot{x} > 0$) and decreases in the lower half-plane ($\dot{x} < 0$). Periodic motion is indicated by a closed curve in the phase-plane (described clockwise). Figure 7.1 shows typical phase-plane solutions for oscillating systems.

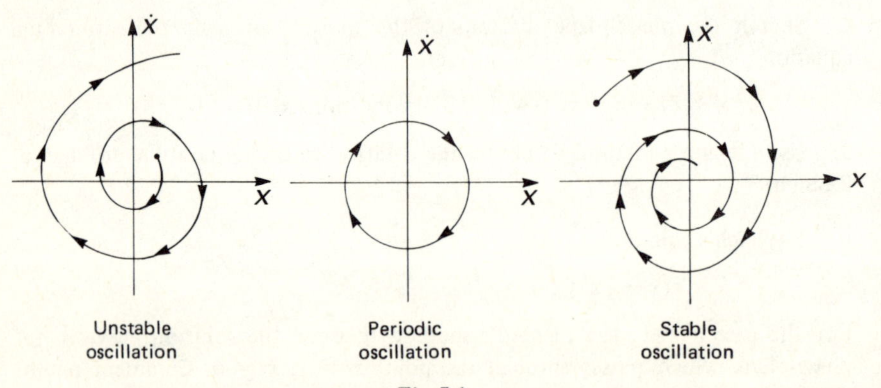

Fig. 7.1

Example 7.1
Solve the simple pendulum equation $\ddot{x} + \sin x = 0$ in the $x\dot{x}$ plane. Find the maximum value of $\dot{x}$ for the cases

(i) $x(0) = -\pi/2, \dot{x}(0) = 0$,

(ii) $x(0) = -\pi, \dot{x}(0) = 0.5$.

Putting $y = \dot{x}$ gives

$$y\frac{dy}{dx} + \sin x = 0 \quad .$$

Therefore $\int y \, dy = -\int \sin x \, dx$.

Therefore $y^2 = 2\cos x + c$,

where the value of c depends on the initial values of x and $\dot{x}$,

i.e. $$\frac{dx}{dt} = \pm\sqrt{2\cos x + c} \quad .$$

It is not possible to solve this differential equation analytically to obtain an xt relationship in terms of simple functions. However, an $x\dot{x}$ graph can be drawn for a particular initial condition, and an xt graph can be obtained approximately following the method given in Section 7.3. However, the maximum value of $\dot{x}$ can be obtained directly from the xy solution.

(i) For $x(0) = -\pi/2$, $\dot{x}(0) = 0$, (i.e. $y = 0$ when $x = -\pi/2$),

$0 = 0 + c$, therefore $c = 0$ and

$y^2 = 2\cos x$.

The maximum value of $\dot{x}$ occurs when $x = 0$,

i.e. $$\dot{x}_{MAX} = y_{MAX} = \sqrt{2} \approx 1.414 \quad .$$

(ii) For $x(0) = -\pi$, $\dot{x}(0) = 0.5$, (i.e. $y = \frac{1}{2}$ when $x = -\pi$),

$0.25 = -2 + c$, therefore $c = 2.25$.

Therefore $y^2 = 2\cos x + 2.25$ and

$$\dot{x}_{MAX} = y_{MAX} = \sqrt{4.25} \approx 2.062 \quad .$$

(Compare with Example 6.2 and Figure 6.2.)

Example 7.2

Figure 7.2 refers to a mass-spring system with friction. Assuming that the friction force is constant for non-zero velocities, the equation of motion is of the form

$$m\ddot{x} = -Kx - F\,\text{sign}\,(\dot{x}) \;,$$

where the function sign $(\dot{x})$ is defined for $\dot{x} \neq 0$ as

$$\text{sign}\,(\dot{x}) = \begin{cases} 1 \;, & \dot{x} > 0 \\ -1 \;, & \dot{x} < 0 \end{cases} .$$

For the case $m = K = F = 1$, $x(0) = 8.5$, $\dot{x}(0) = 0$, show that the motion converges to $x = 0.5, \dot{x} = 0$.

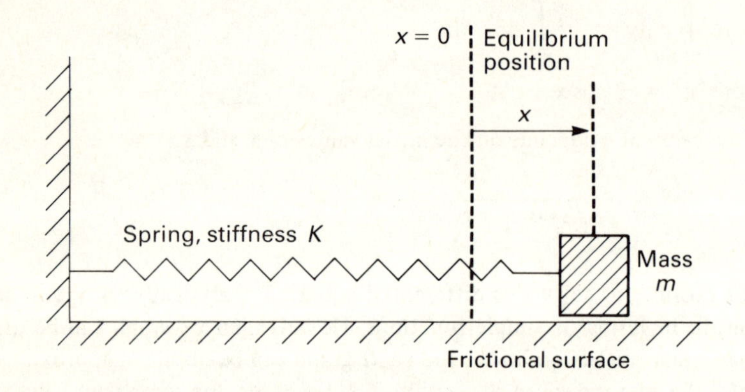

Fig. 7.2

Using the given values for m, K and F, the equation becomes

$$\ddot{x} = -x \mp 1, \text{ for } \dot{x} \gtrless 0 \;.$$

Now put $y = \dot{x}$, to give

$$\int y \,\text{d}y = \int (-x \mp 1)\text{d}x \text{ for } y \gtrless 0 \;,$$

and so $$y^2 = -x^2 \mp 2x + c \text{ for } y \gtrless 0 \;.$$

Therefore $$x^2 + y^2 + 2x = c \text{ for } y > 0 \;,$$

i.e. a family of semicircles centre $(-1,0)$,

and $$x^2 + y^2 - 2x = c \text{ for } y < 0,$$

i.e. a family of semicircles centre $(1,0)$.

Using the given initial point $(8.5, 0)$, and noting that the semicircles must be described clockwise in the phase-plane, it is easy to see that the motion will reach the point $(0.5, 0)$ from a semicircle in the upper half-plane (see Fig. 7.3).

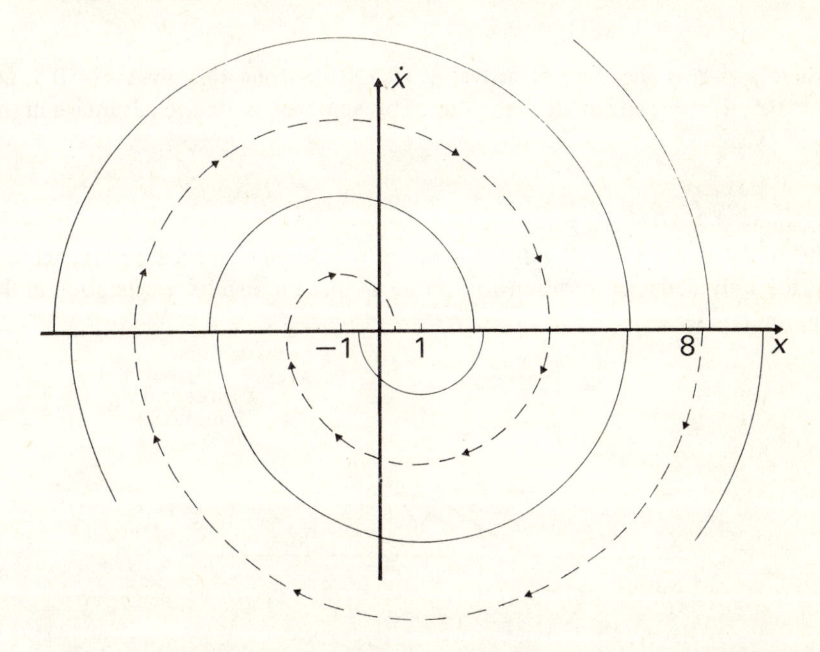

Fig. 7.3 – Phase-plane diagram for $\ddot{x} = -x - \text{sign}(\dot{x})$. Dashed line shows solution curve for $x(0) = 8.5, \dot{x}(0) = 0$.

If we attempt to continue the motion into the lower halfplane, we discover that the point (x, y) arrives on a semicircle whose direction of description immediately returns the point to the upper halfplane. Therefore it appears that the solution is trapped at the point (0.5, 0). We need to look at the physical problem more carefully in order to discover the nature of subsequent motion. The given equation of motion does not apply when $\dot{x} = 0$. For this case, the friction force will equal the applied force until the latter is sufficiently large to cause motion. This will occur when the applied force has magnitude F, i.e. for $\dot{x} = 0$ the friction force has magnitude given by

$$\begin{cases} K|x| \quad , \quad K|x| \leqslant F \\ \\ F \quad\quad , \quad K|x| > F \; . \end{cases}$$

Using the given values for K and F we see that the friction force is balanced by the applied force for $|x| < 1, \dot{x} = 0$, i.e. when the point (x, y) arrives on the x-axis *between the two centres* of the semicircle families. Therefore the point $x = 0.5, \dot{x} = 0$ represents the final position.

An alternative, non-rigorous, but successful approach is to note that the point (x, y) has become trapped on the line $y = 0$, i.e. $\dot{x} = 0$. Therefore any subsequent motion must satisfy the differential equation

$$\dot{x} = 0, \; x(T) = 0.5,$$

where $t = T$ is the time of arrival at $(0.5, 0)$. Solving this gives $x = 0.5$, i.e. $x = 0.5$, $\dot{x} = 0$ is an equilibrium point. This approach is used to advantage in the next example.

Example 7.3

Figure 7.4 shows a certain missile, with fins which can be moved instantaneously to the fully deflected position to provide a constant angular acceleration in the appropriate sense.

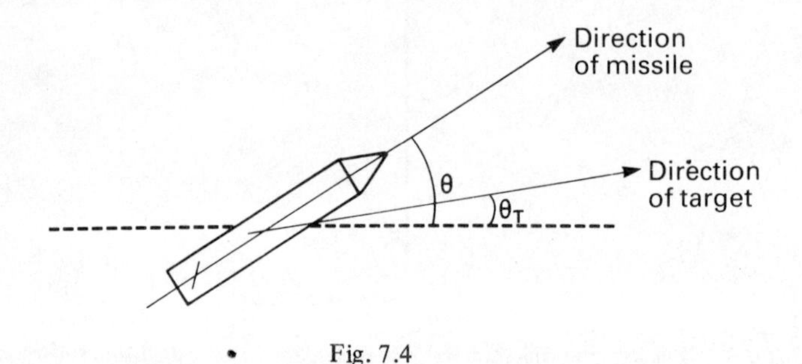

Fig. 7.4

The control equation is

$$\ddot{\theta} = -K \operatorname{sign}(\theta - \theta_T) \;.$$

Sketch the phase plane diagram, and explain why the above control equation is unsatisfactory.

Consider the modified control equation

$$\ddot{\theta} = -K \operatorname{sign}(\theta - \theta_T + k\dot{\theta}) \;,$$

and show that for $k > 0$ the resulting motion in the phase plane is damped.

Put $y = \dot{\theta}$ in the equation $\ddot{\theta} = -K \operatorname{sign}(\theta - \theta_T)$ to give

$$\ddot{\theta} = y\frac{\mathrm{d}y}{\mathrm{d}\theta} \text{ and } \int y \,\mathrm{d}y = \mp \int K \,\mathrm{d}\theta \;\; \text{for } \theta \gtrless \theta_T$$

Thus $y^2 = C - 2K\theta$ for $\theta > \theta_T$,

i.e. a family of parabolas to the *right* of the 'switching line' $\theta = \theta_T$,
and $y^2 = C + 2K\theta$ for $\theta < \theta_T$,
i.e. a family of parabolas to the *left* of the 'switching line' $\theta = \theta_T$.

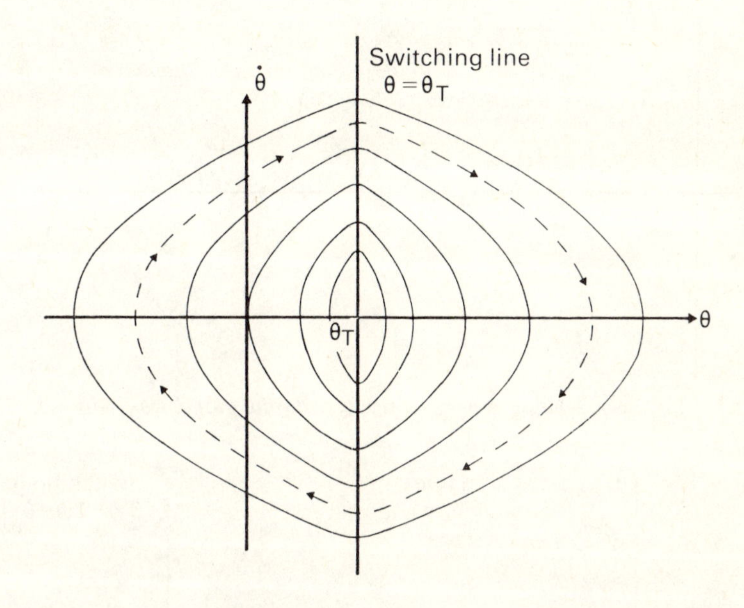

Fig. 7.5 – Phase-plane diagram for $\ddot{\theta} = -K \text{ sign } (\theta - \theta_T)$. Dashed line shows solution curve for $\theta(0) = -\theta_T, \dot{\theta}(0) = 0$.

Figure 7.5 shows the phase-plane diagram, which consists of the two families of parabolas. In particular, the motion arising from an initial point $\theta = -\theta_T$, $y = 0$ is shown, and consists of an oscillation between $\theta = -\theta_T$ and $\theta = 3\,\theta_T$. Figure 7.6 shows the associated θt graph. This is clearly an unsatisfactory form of control, and some form of damping must be provided.

Now consider the modified control equation

$$\ddot{\theta} = -K \text{ sign } (\theta - \theta_T + k\dot{\theta}) \ .$$

The phase-plane diagram consists of the *same* two families of parabolas, but the switching line is now $\theta - \theta_T + ky = 0$. For $k > 0$ the switching line has a negative slope, so that $\ddot{\theta}$ changes sign before θ passes through $\theta = \theta_T$. Figure 7.7 shows typical $\theta\dot{\theta}$ graphs. Starting from the initial point $(-\theta_T, 0)$ we see that successive crossover points on the switching line become closer to the point $(\theta_T, 0)$, until the point $(\theta, \dot{\theta})$ becomes trapped *on* the switching line (in a similar manner to the trapped point in Example 7.2).

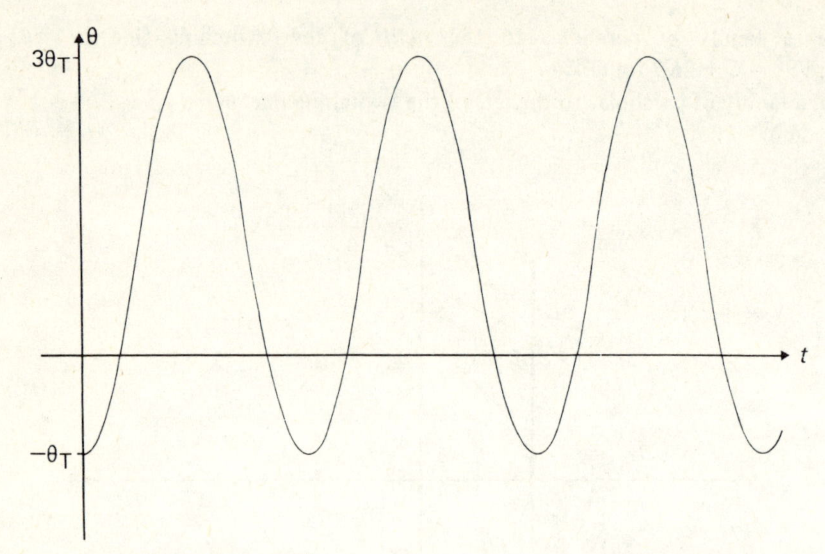

Fig. 7.6 – Solution of $\ddot{\theta} = -K$ sign $(\theta - \theta_T)$ for $\theta(0) = -\theta_T$, $\dot{\theta}(0) = 0$.

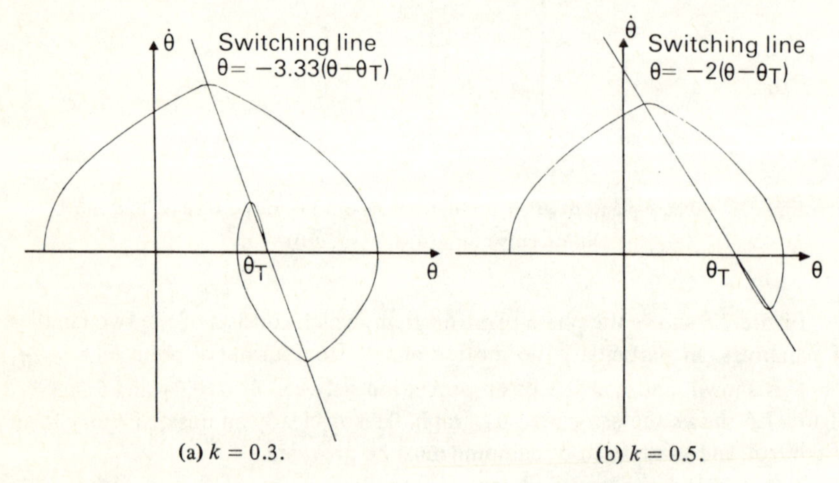

(a) $k = 0.3$. (b) $k = 0.5$.

Fig. 7.7 – Typical phase-plane solutions of $\ddot{\theta} = -K$ sign $(\theta - \theta_T + k\dot{\theta})$.

[The reader is left to prove that the point $(\theta, \dot{\theta})$ becomes trapped on the switching line when θ is such that $|\theta - \theta_T| < K k^2$.]

We can now use the alternative method at the end of Example 7.2 to determine the subsequent motion. If the point (θ, y) is constrained on the line $\theta - \theta_T + ky = 0$, then the motion must satisfy the differential equation

$$k\frac{\mathrm{d}\theta}{\mathrm{d}t} + \theta = \theta_T, \;\; \theta(t_1) = \theta_1 \,,$$

where $t = t_1$ is the time of arrival on the switching line. Solving this gives $\theta = \theta_T + (\theta_1 - \theta_T)\,e^{t_1 - t}$, and we see that θ tends exponentially towards $\theta = \theta_T$. Figure 7.8 shows typical θt graphs. This rather surprising result may be confirmed by constructing a continuous model of the missile control using analog computing elements and a relay switch. However, since the relay is unable to switch instantaneously, the point (θ, y) goes down the switching line towards $(\theta_T, 0)$ in a zigzag fashion, with $\ddot{\theta}$ rapidly changing between $+K$ and $-K$. For a comprehensive treatment of relays, see Jacobs (1974).

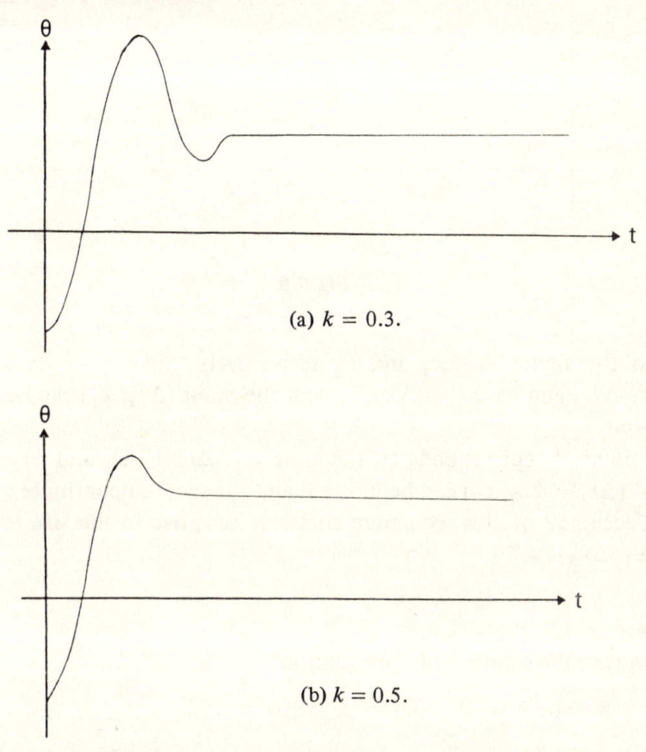

Fig. 7.8 – Typical θt solutions of $\ddot{\theta} = -K \operatorname{sign}(\theta - \theta_T + k\dot{\theta})$.

7.3 GRAPHICAL METHOD FOR *xt* GRAPH FROM PHASE-PLANE SOLUTION

The *xt* solution may be obtained graphically from the phase-plane diagram as follows.

Figure 7.9 shows part of a phase plane solution, with P_0 the initial point ($t = 0$). Let P_1 correspond to the time $t = \Delta t_1$. The change in x can be measured, and an average value of y in the interval $0 < t < \Delta t_1$ can be estimated. These are

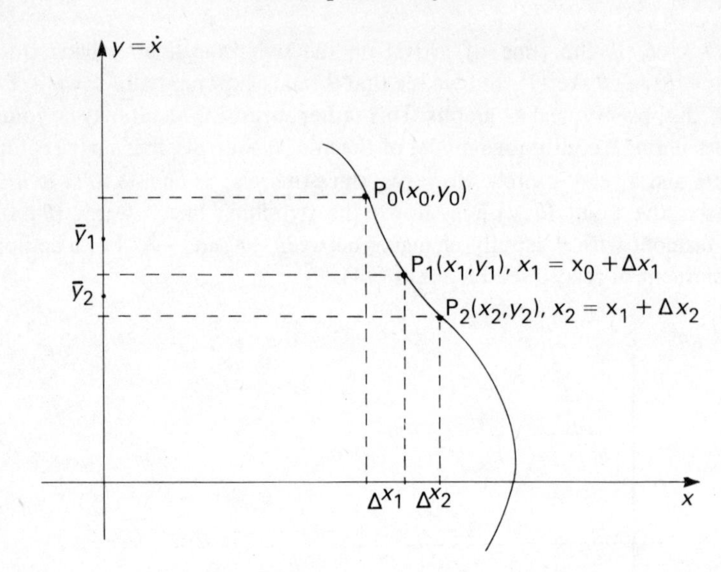

Fig. 7.9

shown in the figure as Δx_1 and $\bar{y}_1$ respectively. Since $y = \mathrm{d}x/\mathrm{d}t$ we have $\bar{y}_1 \approx \Delta x_1/\Delta t_1$, and so $\Delta t_1 \approx \Delta x_1/\bar{y}_1$, and the point $(\Delta t_1, x_1)$ can be plotted on the xt graph.

The point P_2 corresponds to the time $t = \Delta t_1 + \Delta t_2$ and $\Delta t_2 \approx \Delta x_2/\bar{y}_2$. The point $(\Delta t_1 + \Delta t_2, x_2)$ can be plotted, and the method continues.

The accuracy of this technique is clearly sensitive to the size of the increments $\Delta x_1, \Delta x_2, \ldots$.

Problems

1. Show that the solution of the equation

$$\ddot{x} + \omega^2 x = 0, \; x(0) = K, \; \dot{x}(0) = 0,$$

can be represented in the phase-plane by an ellipse. If the phase-plane is modified so that $\dot{x}/\omega$ is plotted against x, show that the curve becomes a circle cente (0,0) radius K. Without considering the solution of the equation, show that the time taken for the point to complete one revolution of the circle is $2\pi/\omega$.

2. Given the differential equation $\ddot{x} + \omega^2 \sin x = 0, x(0) = x_0, \dot{x}(0) = y_0$, show that the maximum of $\dot{x}$ is

$$\sqrt{2\omega^2(1 - \cos x_0) + y_0^2} \; .$$

If $x_0 = -\pi/2$, obtain the condition on the value of y_0 so that $\dot{x} > 0$ for all t.

3. A unit mass moves in a straight line under an attraction $\omega^2 x$ towards the

origin. An additional force, μ, is constant and applied to oppose the motion. The equation of motion is

$$\ddot{x} + \omega^2 \{x + \frac{\mu}{\omega^2} \operatorname{sign}(\dot{x})\} = 0 \; .$$

Assuming that $\omega = 1$, draw the phase-plane solution for the initial conditions $x = 10\mu$, $\dot{x} = 0$. Verify that the mass first comes to rest when $x = -8\mu$, and finally comes to rest after time 5π.

4. Sketch the solution in the phase-plane for the differential equation

$$\ddot{\theta} = -2 \operatorname{sign}(0.5\dot{\theta} + \theta - 0.5), \quad \theta(0) = \tfrac{37}{4}\,, \; \dot{\theta}(0) = 0 \; .$$

Show that the point $(\theta, \dot{\theta})$ becomes constrained on the switching line within $0 < \theta < 1$.

[Obtain the equation of each section of the solution curve.]

5. Draw the family of solution curves in the phase-plane for the equation

$$\frac{d^2x}{dt^2} = f(x)$$

for the cases

(i)

$$f(x) = \begin{cases} a & \text{for} \quad x < -k \\ 0 & \text{for} \quad -k < x < k \\ -a & \text{for} \quad x > k \end{cases} ,$$

(ii)

$$f(x) = \begin{cases} a & \text{for} \quad x < -k \\ -\frac{a}{k}x & \text{for} \quad -k < x < k \\ -a & \text{for} \quad x > k \end{cases} .$$

6. Sketch the phase-plane solution for the equation

$$\ddot{x} + \operatorname{sign}(\dot{x}) = f(x), \quad x(0) = \dot{x}(0) = 0,$$

where $f(x)$ is as in Problem 5(i) with $a = 2, k = 0.5$.

7. Use Liénard's method to confirm the solution for the first half cycle of the equation given in Problem 3. Repeat for different increment sizes.

8. Draw the phase-plane solution for the equation

$$\ddot{x} + x = 0, \quad x(0) = 1, \; \dot{x}(0) = 0 \quad .$$

Apply the graphical technique to obtain the xt graph. Repeat for different increment sizes and compare with the exact solution $x = \cos t$.

9. Given the initial value problem

$$\ddot{x} + 2x - x^2 = 0, \quad x(0) = 1, \; \dot{x}(0) = 0 \quad ,$$

show that the phase-plane solution is

$$y^2 = \tfrac{1}{3}(2x^3 - 6x^2 + 4), \quad \text{where } y = \dot{x} \quad .$$

Deduce that the motion is an oscillation between $x = -(\sqrt{3} - 1)$ and $x = 1$, and state the maximum value of $\dot{x}$.

If the initial conditions are changed to $x(0) = 3, \dot{x}(0) = 0$, show that $x \to \infty$.

Answers to problems

2. $y_0 > \sqrt{2}\,\omega$

9. $\dot{x}_{\text{MAX}} = \dfrac{2}{\sqrt{3}}$.

8

Linearisation techniques

8.1 INTRODUCTION

In this chapter we consider the two dimensional system

$$\begin{aligned} \dot{x} &= P(x,y) \\ \dot{y} &= Q(x,y) \end{aligned} \tag{8.1}$$

where at least one of the functions P, Q is nonlinear.

Note that the second order differential equation

$$\ddot{x} = Q(x,\dot{x})$$

may be written in the form of equation 8.1 by making the substitution $y = \dot{x}$ to obtain

$$\begin{aligned} \dot{x} &= y \\ \dot{y} &= Q(x,y) \quad . \end{aligned} \tag{8.2}$$

The solution of the system (8.1) can be represented by a curve in the xy plane starting from some initial point (x_0,y_0). This xy plane should not be confused with the phase plane as defined in Chapter 7. The term phase plane only applies for the particular case $\dot{x} = y$, e.g. as in the system (8.2).

The solution curves in the xy plane may be obtained approximately in the neighbourhood of *equilibrium points* by a linearisation process, but unlike the linear system, $\dot{\mathbf{x}} = \mathbf{A}\mathbf{x}$, the system (8.1) may have more than one equilibrium point. The linearisation method must be applied near *each* equilibrium point so that a complete picture of the solution curves in the xy plane may be sketched.

8.2 SINGULAR POINTS

We define a SINGULAR POINT in the xy plane for the system (8.1) to be a point where both of the functions $P(x,y)$ and $Q(x,y)$ are zero. Equations 8.1 imply that $\dot{x} = \dot{y} = 0$ at a singular point, so that a singular point is also an equilibrium point. In other words, if (x_1, y_1) is a singular point then a particular solution of equations 8.1 is simply $x = x_1, y = y_1$.

In many practical problems we are interested in the stability of equilibrium points, i.e. if we choose an initial point near (x_1, y_1), does the point (x, y) on the solution curve remain near (x_1, y_1)?

This can be studied by taking a *linear approximation* to the system (8.1) in the neighbourhood of the singular point (x_1, y_1). The linear approximation to equations 8.1 is obtained by expressing $P(x,y)$ and $Q(x,y)$ as Taylor series in the neighbourhood of (x_1, y_1) and neglecting second and higher degree terms. This gives

$$\begin{aligned} \dot{x} &= a(x - x_1) + b(y - y_1) \\ \dot{y} &= c(x - x_1) + d(y - y_1) \end{aligned} \tag{8.3}$$

where the constants a, b, c, d are the values of the partial derivatives at the point (x_1, y_1), i.e.

$$a = \left.\frac{\partial P}{\partial x}\right|_1, \quad b = \left.\frac{\partial P}{\partial y}\right|_1$$

$$c = \left.\frac{\partial Q}{\partial x}\right|_1, \quad d = \left.\frac{\partial Q}{\partial y}\right|_1$$

where the subscript 1 is used to denote evaluation at the point (x_1, y_1).

Substituting $X = x - x_1$, $Y = y - y_1$ and writing in matrix form, equations 8.3 become

$$\begin{bmatrix} \dot{X} \\ \dot{Y} \end{bmatrix} = \begin{bmatrix} a & b \\ c & d \end{bmatrix} \begin{bmatrix} X \\ Y \end{bmatrix} .$$

Assuming distinct eigenvalues we have

$$\begin{bmatrix} X \\ Y \end{bmatrix} = c_1 e^{\lambda_1 t} \mathbf{v}_1 + c_2 e^{\lambda_2 t} \mathbf{v}_2 ,$$

where λ_1, λ_2 are the eigenvalues of the matrix $\begin{bmatrix} a & b \\ c & d \end{bmatrix}$ with $\mathbf{v}_1$ and $\mathbf{v}_2$ as eigenvectors.

So the solution of equations 8.3 is

$$\begin{bmatrix} x \\ y \end{bmatrix} = \begin{bmatrix} x_1 \\ y_1 \end{bmatrix} + c_1 e^{\lambda_1 t} \mathbf{v}_1 + c_2 e^{\lambda_2 t} \mathbf{v}_2 .$$

The stability of this solution depends on both λ_1 and λ_2 being negative, or complex numbers with negative real parts. In these cases $x \to x_1$ and $y \to y_1$ as $t \to \infty$, and the point (x_1, y_1) is a stable equilibrium point. Further the equilibrium point for equation 8.3 can be classified as a SADDLE, NODE, FOCUS, or CENTRE following the methods of Section 3.4. We define the nature of the singular point (x_1,y_1) for the nonlinear system (8.1) to be a saddle, node, focus or centre if the solution curves in the neighbourhood of the point exhibit the appropriate patterns, i.e. similar to those illustrated in Figs. 3.7, 3.8, 3.9 and 3.10. Assuming that $ad - bc \neq 0$ (when second degree terms in the Taylor series become important), we would expect there to be a close correspondence between the nature of equilibrium for the two systems (8.1) and (8.3). Theorem 1 specifies this correspondence.

Theorem 1 (Poincaré's result)
At each singular point of the nonlinear system (8.1) the classification corresponds in both type and stability with the results obtained by considering the linearised system (8.3), with the single exception that a centre for system (8.3) may be either a centre or focus for system (8.1). (In this latter case further study of system (8.1) is necessary, see Example 8.5.)

We recall that the nonlinear system (8.1) may have several singular points and at each one we can carry out the linearisation process and determine the nature and stability of equilibrium. The local solution curves can then be sketched in the neighbourhood of each singular point and the overall xy plane diagram can be built up.

The exceptional case in Poincaré's result causes no problem for the particular type

$$\ddot{x} + f(x) = 0 \ , \qquad (8.4)$$

i.e. a second order differential equation with no term in $\dot{x}$. For this equation Theorem 2 applies.

Theorem 2
If the nonlinear equation $\ddot{x} + f(x) = 0$ has a singular point in the $x\dot{x}$ plane where the linearised equation indicates a centre, then the nonlinear equation also has a centre at this singular point.

Proof
The substitution $y = \dot{x}$ converts the equation $\ddot{x} + f(x) = 0$ to the form of equations 8.1,

i.e.
$$\dot{x} = y \ ,$$
$$\dot{y} = -f(x) \ .$$

Suppose that there is a singular point at $(x_1, 0)$, (i.e. $f(x_1) = 0$), and that the linearised equation indicates a centre, (i.e. $f'(x_1) > 0$).

$$\frac{dy}{dx} = \frac{\dot{y}}{\dot{x}} = -\frac{f(x)}{y} .$$

Therefore $\displaystyle\int y\,dy = \int -f(x)\,dx$.

Therefore $\displaystyle y^2 = -2\int f(x)\,dx + c$

$$= -2F(x) + c,$$

where $F'(x) = f(x)$ and the value of the constant c depends on the choice of initial point.

If we consider the intersection of the solution curve $y^2 = -2F(x) + c$ with the line $x = x_1$ which passes through the singular point $(x_1, 0)$, then the points of intersection are given by

$$y^2 = -2F(x_1) + c .$$

Therefore there cannot be more than two points of intersection. This excludes the possibility of a focus, where the solution curve would spiral around the singular point and intersect the line $x = x_1$ an infinite number of times. Therefore by Theorem 1 the singular point must be a centre.

Theorem 2 simply confirms the intuition that a second order differential equation with *no damping term* cannot exhibit a focus.

Example 8.1

Locate the singular points of the differential equation

$$\ddot{x} + x - 0.25x^2 = 0$$

and determine their nature. Sketch the family of solution curves in the phase-plane.

Put $y = \dot{x}$ to give

$$\dot{x} = y$$

$$\dot{y} = 0.25x^2 - x .$$

Therefore $P = y$ and $Q = 0.25x^2 - x$.

Singular points are where $P = Q = 0$,

i.e. $y = 0,\ x = 0,\ 4$.

Therefore there are *two* singular points (0,0) and (4,0). We now linearise the equation at each singular point in turn; but firstly we require the partial derivatives.

$$\frac{\partial P}{\partial x} = 0, \quad \frac{\partial P}{\partial y} = 1 \ ,$$

$$\frac{\partial Q}{\partial x} = 0.5x - 1 \ , \quad \frac{\partial Q}{\partial y} = 0$$

At the point (0,0) the linearised equations are

$$\begin{bmatrix} \dot{x} \\ \dot{y} \end{bmatrix} = \begin{bmatrix} 0 & 1 \\ -1 & 0 \end{bmatrix} \begin{bmatrix} x \\ y \end{bmatrix}$$

Note that it is not necessary to write down the linearised equations, only the matrix is required.

The eigenvalues are solutions of the equation

$$\begin{vmatrix} -\lambda & 1 \\ -1 & -\lambda \end{vmatrix} = 0 \ ,$$

i.e. $\lambda^2 + 1 = 0$,

i.e. $\lambda = \pm j$.

Therefore the eigenvalues are purely imaginary which implies a *centre* at (0,0). Therefore the nonlinear equation has *either* a centre *or* a focus at (0,0). However, the given equation has no term in $\dot{x}$ and therefore (0,0) is a centre (Theorem 2).

At the point (4,0) the linearised equations are

$$\begin{bmatrix} \dot{x} \\ \dot{y} \end{bmatrix} = \begin{bmatrix} 0 & 1 \\ 1 & 0 \end{bmatrix} \begin{bmatrix} x-4 \\ y \end{bmatrix}$$

and the eigenvalues are obtained from $\begin{vmatrix} -\lambda & 1 \\ 1 & -\lambda \end{vmatrix} = 0$,

i.e. $\lambda^2 - 1 = 0$,

i.e. $\lambda = \pm 1$.

Therefore the eigenvalues are real and opposite sign which implies a saddle at (4,0).

Therefore the nonlinear equation has a saddle at (4,0) (Poincaré's result).

The eigenvectors corresponding to $\lambda = \pm 1$ are

$$\mathbf{v}_1 = \begin{bmatrix} 1 \\ 1 \end{bmatrix} \quad \text{and} \quad \mathbf{v}_2 = \begin{bmatrix} 1 \\ -1 \end{bmatrix}$$

which are the directions of the asymptotes of the family of solution curves near (4,0). Figure 8.1 shows the local solution curves from which the overall xy diagram for the nonlinear equation can be sketched, see Fig. 8.2.

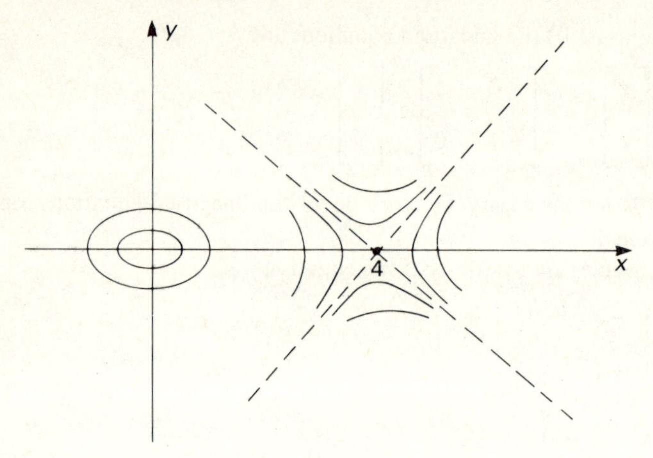

Fig. 8.1 – Solution curves near singular points. The dashed lines show eigenvector directions.

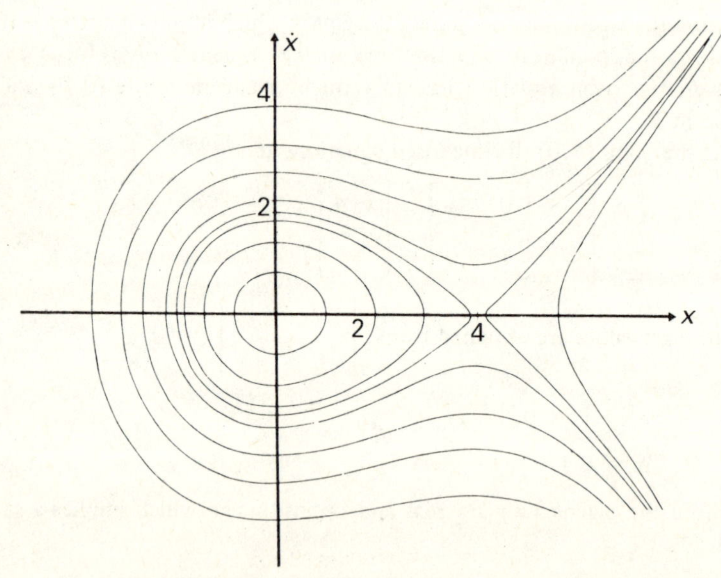

Fig. 8.2 – Solution curves for $\ddot{x} + x - 0.25x^2 = 0$.

Example 8.2

Deduce the following results for the equation in Example 8.1.

(i) The motion will be oscillatory only when the initial point $(x_0, \dot{x}_0)$ lies inside the loop of the curve

$$6(x^2 + y^2) - x^3 = 32 .$$

(ii) For the initial conditions $x(0) = 3$, $\dot{x}(0) = 0$ the oscillatory motion is such that $-\frac{1}{2}(\sqrt{45} - 3) \leqslant x \leqslant 3$ and $|\dot{x}| \leqslant 3\sqrt{2}/2$.

(i) $$\frac{dy}{dx} = \frac{\dot{y}}{\dot{x}} = \frac{0.25x^2 - x}{y} .$$

Therefore $\int y \, dy = \int (0.25x^2 - x) \, dx$.

Therefore $y^2 = x^3/6 - x^2 + c$.

The largest closed curve in the phase-plane passes just inside (4, 0).

Therefore take (4, 0) as initial point to give this curve,

$$0 = \tfrac{32}{3} - 16 + c, \qquad \text{i.e. } c = \tfrac{16}{3} ,$$

and so $$y^2 = x^3/6 - x^2 + \tfrac{16}{3} ,$$

i.e. $$6(x^2 + y^2) - x^3 = 32.$$

Motion will be oscillatory provided initial point lies inside the loop of this curve. The curve is called the SEPARATRIX; it separates the oscillatory and unstable motions (see Fig. 8.3).

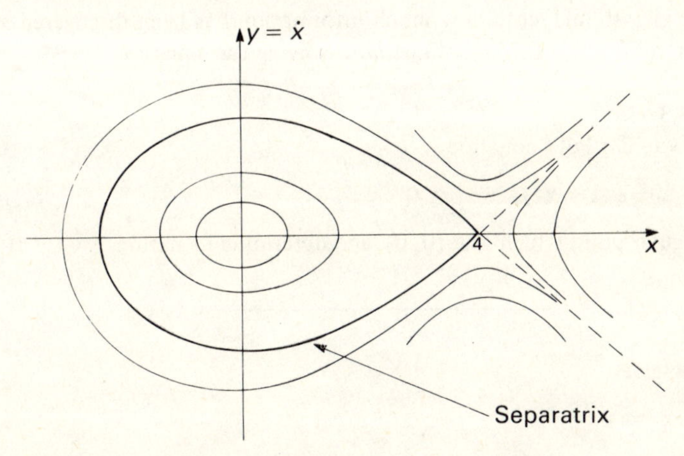

Fig. 8.3 – Separatrix for $\ddot{x} + x - 0.25x^2 = 0$

(ii) As in (i), $y^2 = x^3/6 - x^2 + c$.

The initial point is $(3, 0)$.

Therefore $\quad 0 = 4.5 - 9 + c$

Therefore $\quad c = 4.5$ and solution curve is

$$y^2 = x^3/6 - x^2 + 4.5,$$

$$\text{i.e.} \quad 6y^2 = x^3 - 6x^2 + 27.$$

$(x-3)$ is a factor of the RHS and so $6y^2 = (x-3)(x^2 - 3x - 9)$.
Therefore curve cuts x-axis again when $x^2 - 3x - 9 = 0$,

$$\text{i.e.} \qquad x = \frac{3 \pm \sqrt{9+36}}{2},$$

i.e. $\qquad x = \dfrac{-(\sqrt{45} - 3)}{2}$ is where the *closed* curve cuts the x-axis again.

Therefore the oscillatory motion is such that $-\frac{1}{2}(\sqrt{45} - 3) \leqslant x \leqslant 3$.
The maximum (and minimum) value of $\dot{x} = y$ is when $x = 0$, $(\mathrm{d}y/\mathrm{d}x = 0)$.

Therefore $\quad y^2 = 4.5$,

$$\text{i.e.} \qquad y = \pm\sqrt{4.5} = \pm\frac{3\sqrt{2}}{2}.$$

Therefore the oscillation is such that $|\dot{x}| \leqslant \dfrac{3\sqrt{2}}{2}$.

The reader should note how much information has been discovered about the solution of $\ddot{x} + x - 0.25x^2 = 0$ *without* solving the equation.

Example 8.3

Show that Van der Pol's equation

$$\ddot{x} - \epsilon(1 - x^2)\dot{x} + x = 0$$

has one singular point which is at $(0, 0)$, and determine its nature.

Put $y = \dot{x}$ to give

$$\dot{x} = y = P$$

$$\dot{y} = \epsilon(1 - x^2)y - x = Q .$$

$P = Q = 0$ only when $y = 0, x = 0$.

Therefore (0,0) is the only singular point.

The partial derivatives are

$$\frac{\partial P}{\partial x} = 0, \quad \frac{\partial P}{\partial y} = 1 ,$$

$$\frac{\partial Q}{\partial x} = -2\epsilon xy - 1 , \quad \frac{\partial Q}{\partial y} = \epsilon(1 - x^2) ,$$

and at the point (0,0) the linearised equations are

$$\begin{bmatrix} \dot{x} \\ \dot{y} \end{bmatrix} = \begin{bmatrix} 0 & 1 \\ -1 & \epsilon \end{bmatrix} \begin{bmatrix} x \\ y \end{bmatrix} .$$

The eigenvalues are the solutions of

$$\begin{vmatrix} -\lambda & 1 \\ -1 & \epsilon - \lambda \end{vmatrix} = 0 ,$$

i.e. $\lambda^2 - \epsilon\lambda + 1 = 0$,

i.e. $\lambda = \dfrac{\epsilon \pm \sqrt{\epsilon^2 - 4}}{2}$.

There are several cases to consider.

$\epsilon \leqslant -2$, both eigenvalues are real and negative, therefore (0,0) is a *stable node.*

$-2 < \epsilon < 0$, eigenvalues are complex with negative real parts, therefore (0,0) is a *stable focus.*

$\epsilon = 0$, eigenvalues are purely imaginary, therefore *centre.*

$0 < \epsilon < 2$, eigenvalues are complex with positive real parts, therefore *unstable focus.*

$\epsilon \geqslant 2$, both eigenvalues are real and positive, therefore *unstable node.*

These results apply also to nonlinear equation by Poincaré's result. (The doubtful centre is confirmed simply by noting that the equation is linear when $\epsilon = 0$, i.e. $\ddot{x} + x = 0$.)

It should be noted that the linearised equations only predict the nonlinear solution *near* the singular point. Figures 8.4(a) and 8.4(b) show typical solution curves for the cases $\epsilon = 1$, $\epsilon = 3$. Note that the final motion is a particular oscillation (limit cycle) independent of initial point. Limit cycles are considered in Part III.

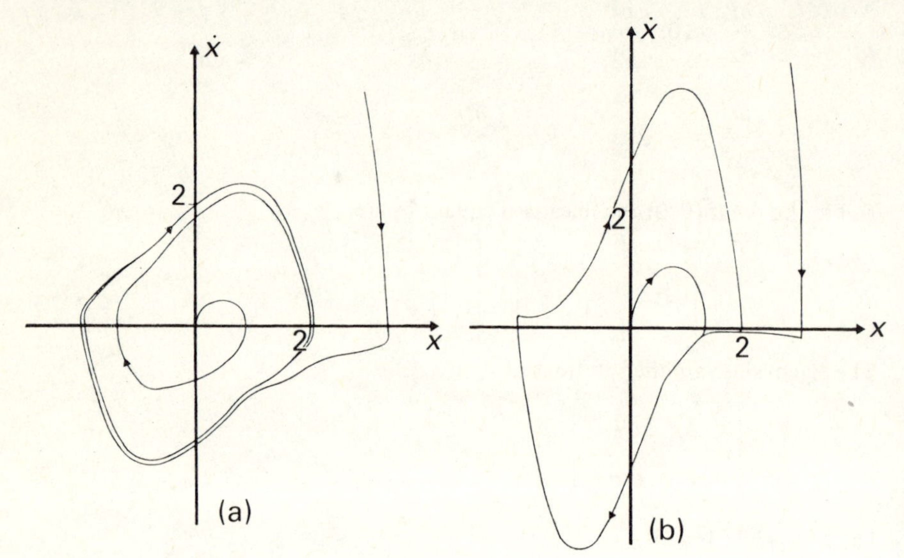

Fig. 8.4 – (a) Solution curves for $\ddot{x} - (1 - x^2)\dot{x} + x = 0$.
(b) Solution curves for $\ddot{x} - 3(1 - x^2)\dot{x} + x = 0$.

Example 8.4

Find the singular points of the equation

$$\ddot{x} + x - x^2 - 2x^3 = 0$$

and determine their nature.

Sketch the family of curves in the phase-plane.

Given that $\dot{x}(0) = 0$, show that the motion will be oscillatory provided

$$\frac{-(5-\sqrt{10})}{6} < x(0) < \frac{1}{2}\,.$$

Put $y = \dot{x}$ to give

$$\dot{x} = y = P,$$

$$\dot{y} = 2x^3 + x^2 - x = Q\,.$$

$P = Q = 0$ when $y = 0$ and $x(2x^2 + x - 1) = 0$,

i.e. $\qquad x(2x-1)(x+1) = 0\,.$

Therefore singular points at (0,0), (−1,0), (0.5,0).

The partial derivatives are

$$\frac{\partial P}{\partial x} = 0 \ , \quad \frac{\partial P}{\partial y} = 1 \ ,$$

$$\frac{\partial Q}{\partial x} = 6x^2 + 2x - 1 \ , \quad \frac{\partial Q}{\partial y} = 0 \ .$$

At the point (0,0) the linearised equations are

$$\begin{bmatrix} \dot{x} \\ \dot{y} \end{bmatrix} = \begin{bmatrix} 0 & 1 \\ -1 & 0 \end{bmatrix} \begin{bmatrix} x \\ y \end{bmatrix} .$$

The eigenvalues are $\pm j$ which implies a *centre*, confirmed for the nonlinear equation by Theorem 2.

At the point (−1,0) the linearised equations are

$$\begin{bmatrix} \dot{x} \\ \dot{y} \end{bmatrix} = \begin{bmatrix} 0 & 1 \\ 3 & 0 \end{bmatrix} \begin{bmatrix} x+1 \\ y \end{bmatrix} .$$

The eigenvalues are $\pm \sqrt{3}$ which implies a *saddle*, confirmed for nonlinear equation by Poincaré's result.

At the point (0.5,0) the linearised equations are

$$\begin{bmatrix} \dot{x} \\ \dot{y} \end{bmatrix} = \begin{bmatrix} 0 & 1 \\ 3/2 & 0 \end{bmatrix} \begin{bmatrix} x-0.5 \\ y \end{bmatrix}$$

The eigenvalues are $\pm \sqrt{3/2}$ which implies a *saddle*, confirmed for nonlinear equation by Poincaré's result.

The phase-plane diagram in the neighbourhood of singular points is shown in Fig. 8.5, and Fig. 8.6 shows the completed phase-plane diagram.

In drawing Fig. 8.6 we have assumed that the separatrix passes through the singular point (0.5,0). This is confirmed below.

$$\frac{\mathrm{d}y}{\mathrm{d}x} = \frac{\dot{y}}{\dot{x}} = \frac{2x^3 + x^2 - x}{y}$$

Therefore $\displaystyle\int y \,\mathrm{d}y = \int (2x^3 + x^2 - x)\,\mathrm{d}x$.

Therefore $y^2 = x^4 + \dfrac{2x^3}{3} - x^2 + c$.

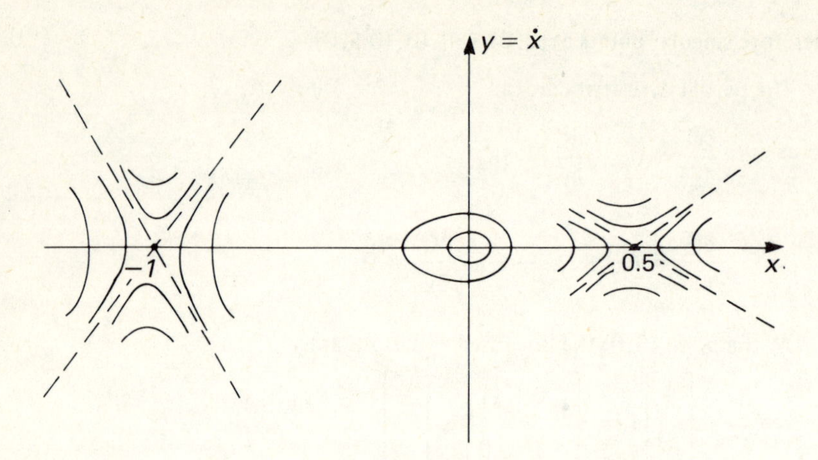

Fig. 8.5 – Solution curves near singular points.

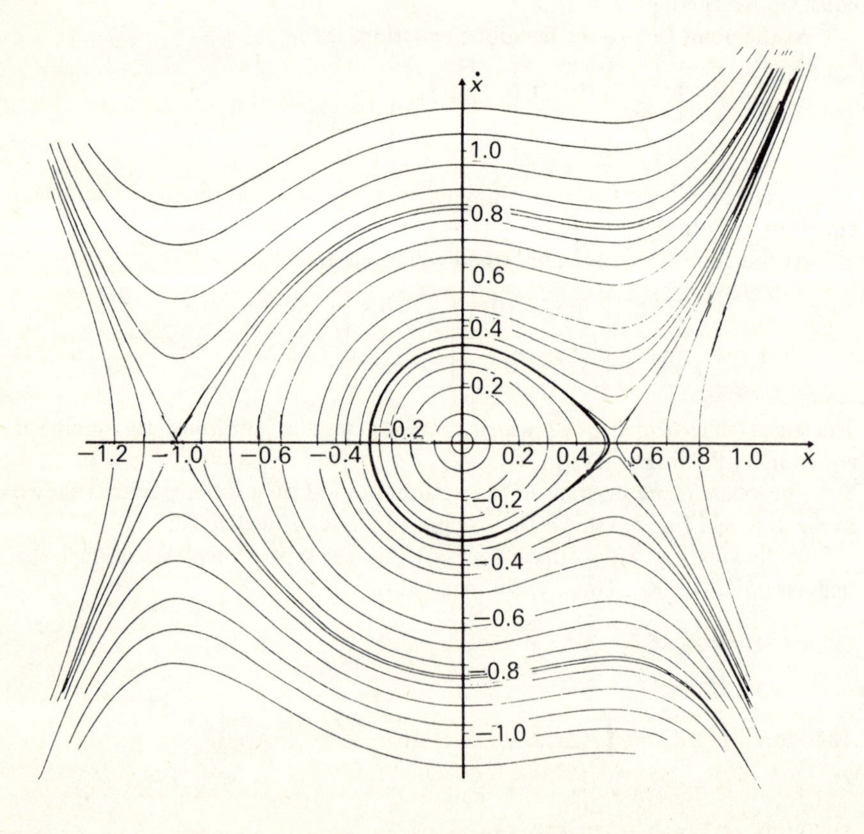

Fig. 8.6 – Solution curves for $\ddot{x} + x - x^2 - 2x^3 = 0$

Taking (0.5,0) as a point on the separatrix we get

$$0 = \tfrac{1}{16} + \tfrac{1}{12} - \tfrac{1}{4} + c \ .$$

Therefore $c = \tfrac{5}{48}$

and $y^2 = \tfrac{1}{48}(48x^4 + 32x^3 - 48x^2 + 5)$ is the equation of the separatrix. $(2x - 1)$ is a double factor of the function on the right-hand side

and so $\quad y^2 = \tfrac{1}{48}(2x-1)^2\ (12x^2 + 20x + 5)$.

Therefore separatrix cuts x-axis again when

$$12x^2 + 20x + 5 = 0 \ ,$$

$$\text{i.e. } x = \frac{-20 \pm \sqrt{400 - 240}}{24} = \frac{-5 \pm \sqrt{10}}{6} \ .$$

The value between -1 and 0.5 is $\dfrac{-5 + \sqrt{10}}{6}$.

The initial point must lie inside loop of separatrix for oscillatory motion.

If $\dot{x}(0) = y(0) = 0$ then $\dfrac{-(5 - \sqrt{10})}{6} < x(0) < \dfrac{1}{2}$ is the condition for oscillatory motion.

Note that if we had assumed that $(-1,0)$ was on separatrix we would have found no further intersection of x-axis between -1 and 0.5.

Example 8.5

The pair of differential equations

$$\dot{x} = 0.5x - xy$$

$$\dot{y} = -2y + xy \ , \quad x, y \geqslant 0 \ ,$$

occur in a study of interacting populations. Find the singular points and determine their nature. Sketch the family of solution curves in the first quadrant of the xy plane.

$$P = 0.5\,x - xy \ , \quad Q = -2y + xy \ .$$

$$P = x(0.5 - y) = 0 \text{ when } x = 0 \text{ or } y = 0.5,$$

$$Q = y(-2 + x) = 0 \text{ when } x = 2 \text{ or } y = 0 \ .$$

Therefore $P = Q = 0$ at $(0,0)$ and $(2,0.5)$ which are the singular points.

$$\frac{\partial P}{\partial x} = 0.5 - y \; , \quad \frac{\partial P}{\partial y} = -x \; ,$$

$$\frac{\partial Q}{\partial x} = y \; , \quad \frac{\partial Q}{\partial y} = -2 + x \; .$$

At $(0,0)$ the linearised equations are

$$\begin{bmatrix} \dot{x} \\ \dot{y} \end{bmatrix} = \begin{bmatrix} 0.5 & 0 \\ 0 & -2 \end{bmatrix} \begin{bmatrix} x \\ y \end{bmatrix} .$$

Therefore the eigenvalues are 0.5, −2 which implies a *saddle,* confirmed for nonlinear equations.

Note that the eigenvectors are $\begin{bmatrix} 1 \\ 0 \end{bmatrix}$ and $\begin{bmatrix} 0 \\ 1 \end{bmatrix}$, which correspond to the directions of the coordinate axis.

At $(2,0.5)$ the linearised equations are

$$\begin{bmatrix} \dot{x} \\ \dot{y} \end{bmatrix} = \begin{bmatrix} 0 & -2 \\ 0.5 & 0 \end{bmatrix} \begin{bmatrix} x-2 \\ y-0.5 \end{bmatrix} .$$

Therefore the eigenvalues are solutions of $\lambda^2 + 1 = 0$, i.e. $\lambda = \pm j$, which implies a *centre.*

Therefore the nonlinear system has *either* a centre *or* focus at $(2,0.5)$. It is not obvious by inspection of the nonlinear equations which type of singular point we have. The following analysis confirms that it *is* a centre.

$$\frac{\mathrm{d}y}{\mathrm{d}x} = \frac{\dot{y}}{\dot{x}} = \frac{y(x-2)}{x(0.5-y)}$$

Therefore $$\int \frac{(0.5-y)}{y}\,\mathrm{d}y = \int \frac{(x-2)}{x}\,\mathrm{d}x \; ,$$

and so $$0.5 \ln y - y = x - 2 \ln x + c \; .$$

The equation of the solution curve which passes through the point $(2, 1)$ is

$$0.5 \ln y - y = x - 2 \ln x - 3 + 2 \ln 2 \; .$$

Consider the intersection of this curve with the line $x = 2$ which passes through the singular point. If (2, 0.5) is a centre there will be just two intersections, otherwise there will be an infinite number. The curve intersects when

$$0.5 \ln y - y = -1 \quad ,$$

i.e. $y = e^{2(y-1)}$, which has just two solutions, see Fig. 8.7.

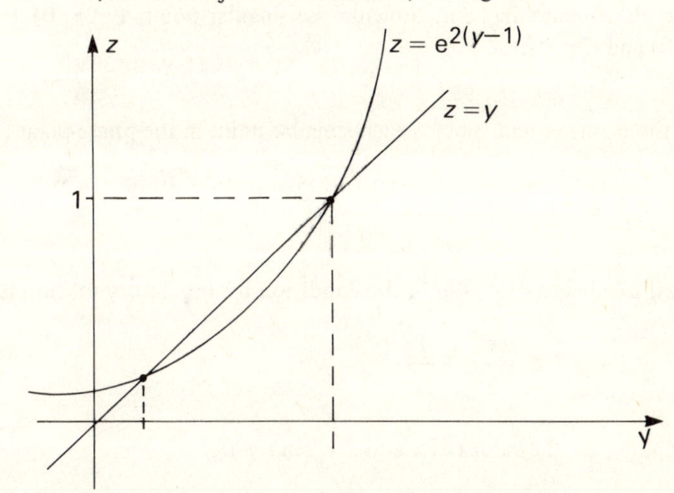

Fig. 8.7 – Two solutions for $y = e^{2(y-1)}$

The solution curves in the first quadrant of the xy plane are shown in Fig. 8.8.

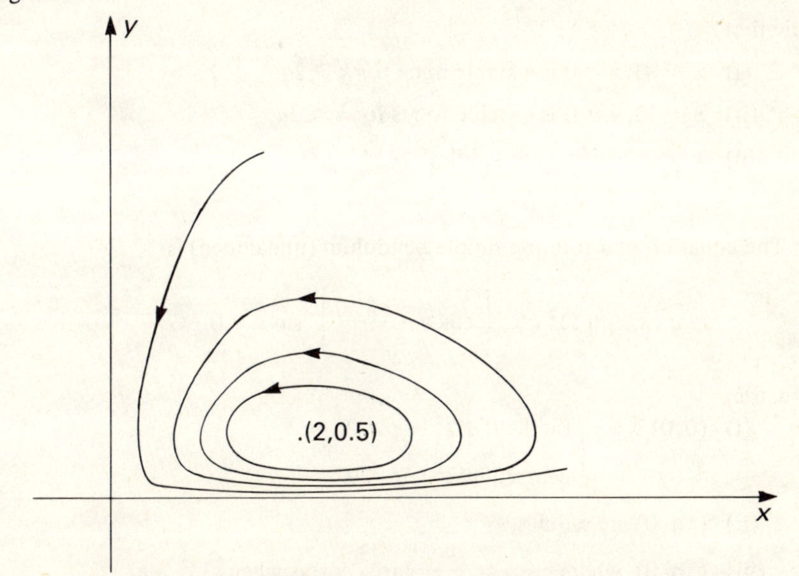

Fig. 8.8 – Solution curves for $\dot{x} = 0.5x - xy$, $\dot{y} = -2y + xy$, $x, y \geqslant 0$

Problems

1. Show that the pendulum equation

$$\ddot{x} + \sin x = 0$$

has centres at $(2n\pi, 0)$ and saddles at $(2(n+1)\pi, 0)$, $n = 0, \pm 1, \pm 2, \ldots$. Sketch the phase-plane diagram, showing the singular points $(-2\pi, 0)$, $(-\pi, 0)$, $(0,0)$, $(\pi, 0)$ and $(2\pi, 0)$.

2. Find the position and type of each singular point in the phase-plane, for the equation

$$\ddot{x} + x - \mu x^3 = 0 \ .$$

Show that if $\mu > 0$ and $|x_0| < \mu^{-\frac{1}{2}}$, the condition for oscillatory motion is

$$x_0^2 + y_0^2 < \frac{\mu x_0^4}{2} + \frac{1}{2\mu} \ ,$$

where x_0 and y_0 are the values of x and $\dot{x}$ when $t = 0$.

3. The equation of a damped pendulum is

$$\ddot{x} + k\dot{x} + n^2 \sin x = 0, \ k \text{ positive}.$$

Show that

(i) $x = 0, \dot{x} = 0$ is a stable node for $k \geqslant 2n$,

(ii) $x = 0, \dot{x} = 0$ is a stable focus for $k < 2n$,

(iii) $x = \pm\pi, \dot{x} = 0$ are saddles.

4. The equation of a rotating simple pendulum (undamped) is

$$m\,a^2\,\ddot{x} - m\,\Omega^2\,a^2 \left\{ \cos x - \frac{g}{a\,\Omega^2} \right\} \sin x = 0 \ .$$

Show that

(i) $(0,0)$ is a $\begin{cases} \text{saddle if } a\Omega^2 - g > 0 \\ \text{centre if } a\Omega^2 - g < 0 \end{cases}$,

(ii) $(\pm\pi, 0)$ are saddles,

(iii) $(\pm\alpha, 0)$, where $\cos\alpha = \dfrac{g}{a\Omega^2}$, are centres when $a\Omega^2 > g$.

5. Find the position of singular points, their type and stability for the nonlinear system

$$\dot{x} = -x + xy\,,$$

$$\dot{y} = 2y - xy + 0.5x\ .$$

6. Show that the nonlinear system

$$\dot{x} = 0.5x - xy + \alpha(x + y)$$

$$\dot{y} = -2y + xy - \alpha(x + y), \quad |\alpha| < 0.1\ ,$$

has a saddle at $(0,0)$ and a focus at $(2 + 5\alpha, (2 + 5\alpha)/4)$. Show that the focus is stable for $\alpha < 0$ and unstable for $\alpha > 0$.

[The case $\alpha = 0$ has been studied in Example 8.5.]

7. (i) Show that the singular points of the differential equation

$$\ddot{x} + g(x) = 0\ ,$$

where $g(x)$ is a polynomial in x, all lie on the x axis of the $x\dot{x}$ plane and can only be centres or saddles. State a condition on the value of $g'(x)$ at a singular point which determines whether the point is a centre or a saddle. Deduce that centres and saddles occur alternately on the x axis.

(ii) Confirm that the differential equation

$$\ddot{x} + x - x^3 = 0$$

has a centre at $(0,0)$ and saddles at $(-1,0)$ and $(1,0)$. Find the equation of the separatrix and, given that $x(0) = 0$, obtain a condition on the value of $\dot{x}(0)$ so that the solution is oscillatory.

8. Show that the equation

$$\ddot{x} + x = 0.1\,(x^2 + 2x^3)$$

has three singular points which are at $(0,0)$, $(-2.5,0)$, $(2,0)$ in the phase-plane. Confirm that one of these is a centre and the other two are saddles, and sketch the phase-plane diagram you would expect. Given that $\dot{x}(0) = 0$, find the condition on $x(0)$ for the motion to be oscillatory.

9. A nonlinear system is described by the differential equations

$$\dot{x} = kx - y^2\ ,$$

$$\dot{y} = -y + x^2\ .$$

Show that the system has singular points at $(0,0)$ and $(k^{\frac{1}{3}}, k^{\frac{2}{3}})$.
Show that no stable solution can exist when $k > 1$.

10. Given the initial value problem

$$\ddot{x} + x - x^2 = -2,\ x(0) = u,\ \dot{x}(0) = 0,$$

show that oscillatory motion will result only if $-\frac{5}{2} < u < 2$.

11. Given that $f(y)$ is a polynomial in y such that $f'(0) \neq 0$, show that the differential equation

$$\ddot{x} + x + \epsilon f(\dot{x}) = 0,\ \epsilon \neq 0\ ,$$

has only one singular point, which is at $(-\epsilon f(0), 0)$ in the phase-plane, and that this singular point can only be a focus or a node. Write down a condition for its stability.
In particular, for $f(y) = y^3/3 - y$ and $0 < \epsilon < 2$, state the position, type and stability of the singular point.

12. A certain relaxation oscillator is described by the differential equation

$$\ddot{x} + 2\alpha\dot{x} - \frac{\beta^2}{2}x + x^3 = 0,\ \alpha,\ \beta > 0\ .$$

Show that the singularity at $x = \beta/\sqrt{2}$, $\dot{x} = 0$ is a stable focus if $\alpha < \beta$ and a stable node if $\alpha > \beta$.

Answers to problems

1.

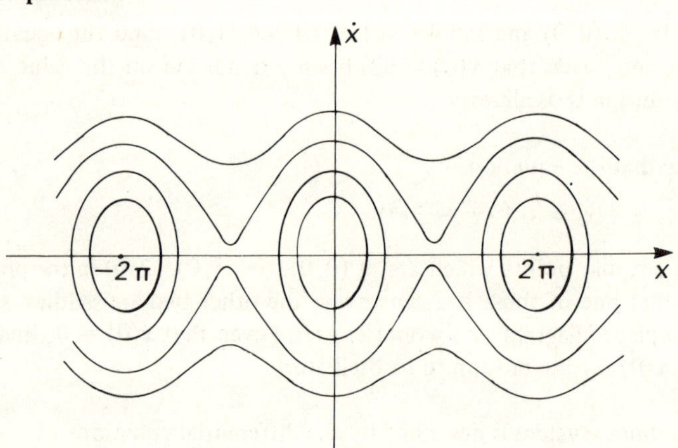

2. $\mu > 0$; centre at $(0,0)$, saddles at $\left(\frac{\pm 1}{\sqrt{\mu}}, 0\right)$;

$\mu < 0$; centre at $(0,0)$.

5. (0,0) saddle, (4, 1) stable focus.

7. (i) $g'(x) > 0$ for centre, $g'(x) < 0$ for saddle.

   (ii) $x^2 + y^2 = \frac{1}{2}x^4 + \frac{1}{2}$, $|\dot{x}(0)| < \dfrac{1}{\sqrt{2}}$

8.

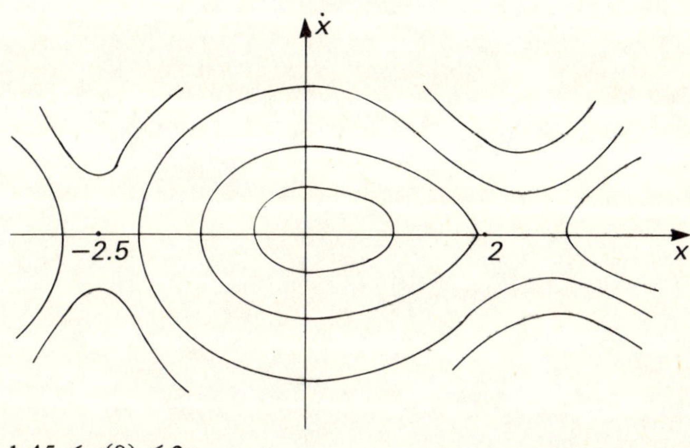

$-1.45 < x(0) < 2.$

11. $\epsilon f'(0) > 0$, (0,0) is unstable focus.

Part III

Nonlinear differential equations – asymptotic methods

9

The methods of Poincaré and Lindstedt

9.1 INTRODUCTION

We have seen in Part II how a small nonlinear term in an equation changes the solution away from the linear one. Sometimes this change is quite dramatic, as in the examples in Chapter 5, but usually a small nonlinear term results in a small change away from the linear solution.

We often encounter small nonlinearities in dynamical systems – that is, physical systems which can be modelled by a differential equation – such as small departures from Hooke's law for springs, small departures from linear voltage characteristics of a diode, and so on. If we measure this smallness by ϵ, it is reasonable to suppose that the outcome will be roughly what it was with $\epsilon = 0$ plus some correction terms which will clearly depend on ϵ.

To illustrate the essential ideas, consider the algebraic equation

$$x = 1 + \epsilon x^3 \;.$$

This cubic equation, which can be written

$$x^3 - \frac{1}{\epsilon}x + \frac{1}{\epsilon} = 0 \;,$$

can be shown to have three real roots for $0 < \epsilon < 0.148$. Obviously one of these roots will be near $x = 1$ – the solid line on Fig. 9.1 shows this root plotted against ϵ. It is clear that $x \to 1$ as $\epsilon \to 0$, and that the correction *does* depend on ϵ.

To copy this behaviour, we set $x = 1 + \epsilon x_1 + \epsilon^2 x_2 + \ldots$, where the x_n's are numbers which we will determine. Substituting into the equation gives

$$\begin{aligned} &1 + \epsilon x_1 + \epsilon^2 x_2 + \ldots \\ &= 1 + \epsilon(1 + \epsilon x_1 + \epsilon^2 x_2 + \ \ldots)^3 \\ &= 1 + \epsilon(1 + 3\epsilon x_1 + 3\epsilon^2 x_2 + 3\epsilon^2 x_1{}^2 + \ \ldots) \;. \end{aligned}$$

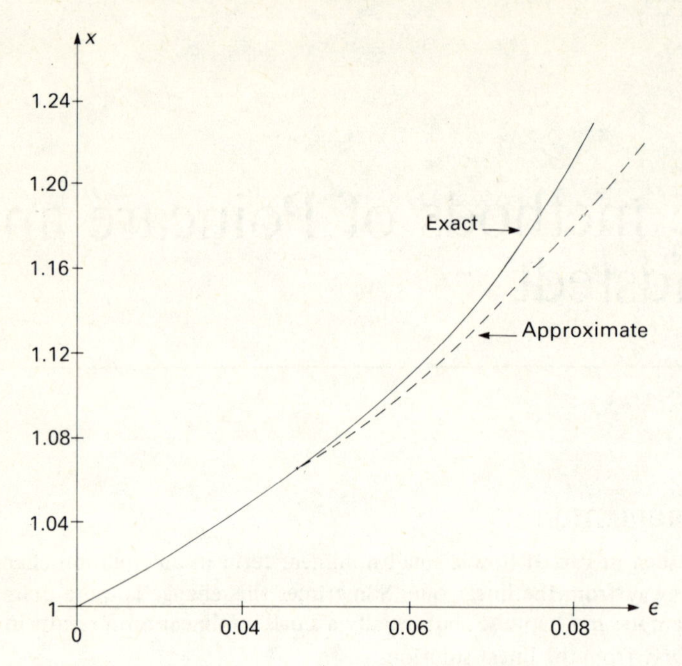

Fig. 9.1 – Solution of $x = 1 + \epsilon x^3$ for various values of ϵ

Since this must be true for *any* value of ϵ, the coefficients of ϵ^n must be equal on the left-hand side and right-hand side. Using the symbol $\lfloor \epsilon^n$ to indicate that we are equating the coefficients of ϵ^n, we have the following.

$$\lfloor \epsilon^\circ : \quad 1 = 1$$

$$\lfloor \epsilon : \quad x_1 = 1$$

$$\lfloor \epsilon^2 : \quad x_2 = 3x_1 = 3$$

$$\lfloor \epsilon^3 : \quad x_3 = 3x_2 + 3{x_1}^2 = 12$$

etc.

Thus $x = 1 + \epsilon + 3\epsilon^2 + 12\epsilon^3 + \ldots$, which is shown as the dashed line on Fig. 9.1. We can see that the cubic expansion is excellent up to about $\epsilon = 0.06$.

9.2 POINCARÉ'S METHOD

Eventually we want to apply these ideas to oscillating systems (as in the previous chapter), but firstly let us consider the first order ordinary differential equation (ODE)

$$\dot{x} + x = 1 + \epsilon x^3, \text{ with } x(0) = 0 \; .$$

The first thing to notice, of course, is that if $\epsilon = 0$ we need no new method to help us solve the equation. For the case $\epsilon = 0$ we have

$$\dot{x} + x = 1, \text{ with } x(0) = 0 ,$$

which may be solved by separating the variables or using the integrating factor method to give

$$x(t) = 1 - \mathrm{e}^{-t} .$$

This will provide a useful check on our new method.

To use the asymptotic method we now set

$$x = x_0 + \epsilon x_1 + \epsilon^2 x_2 + \ldots ,$$

where $x_n = x_n(t)$, and substitute into the original equation. The ODE becomes

$$(\dot{x}_0 + \epsilon \dot{x}_1 + \epsilon^2 \dot{x}_2 + \ldots) + (x_0 + \epsilon x_1 + \epsilon^2 x_2 + \ldots)$$
$$= 1 + \epsilon (x_0^3 + 3\epsilon x_0^2 x_1 + \ldots) ,$$

and the initial condition becomes

$$x_n(0) = 0 \text{ for all } n .$$

Equating similar powers of ϵ on both sides of the equation now gives the following.

$$\underline{|\epsilon^0} : \quad \dot{x}_0 + x_0 = 1$$

$$\underline{|\epsilon} : \quad \dot{x}_1 + x_1 = x_0^3$$

$$\underline{|\epsilon^2} : \quad \dot{x}_2 + x_2 = 3x_0^2 x_1$$

and so on.

From the first equation we obtain $x_0 = 1 - \mathrm{e}^{-t}$, the known solution for the case $\epsilon = 0$.

From the second equation, $\dot{x}_1 + x_1 = 1 - 3\mathrm{e}^{-t} + 3\mathrm{e}^{-2t} - \mathrm{e}^{-3t}$.
This may be solved using Laplace transforms, for instance.

$$(s+1)\, X_1(s) = \frac{1}{s} - \frac{3}{s+1} + \frac{3}{s+2} - \frac{1}{s+3} .$$

Therefore $$X_1(s) = \left(\frac{1}{s} - \frac{1}{s+1}\right) - \frac{3}{(s+1)^2} + \left(\frac{3}{s+1} - \frac{3}{s+2}\right)$$
$$- \frac{1/2}{s+1} - \frac{1/2}{s+3} .$$

Therefore $$x_1(t) = 1 + \tfrac{3}{2}\,\mathrm{e}^{-t} - 3t\,\mathrm{e}^{-t} - 3\mathrm{e}^{-2t} + \tfrac{1}{2}\,\mathrm{e}^{-3t} .$$

[Check: $x_1(0) = 0$.]

Thus $\qquad x(t) = 1 - e^{-t} + \epsilon(1 + \frac{3}{2}e^{-t} - 3te^{-t} - 3e^{-2t} + \frac{1}{2}e^{-3t}) + \ldots$.

Graphing this for various values of ϵ shows that the approximation is good for small ϵ *until* t gets large. This is shown in Fig. 9.2.

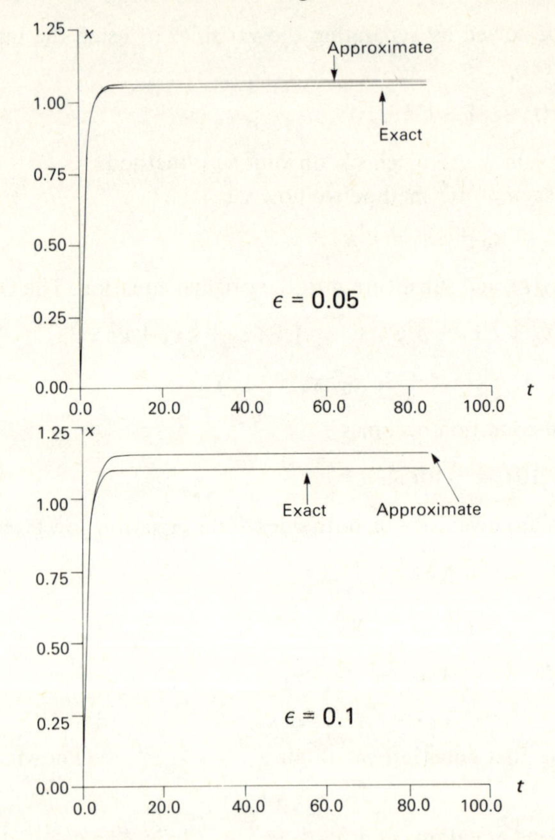

Fig. 9.2 – Solutions of $\dot{x} + x = 1 + \epsilon x^3$, with $x(0) = 0$, for $\epsilon = 0.05, 0.1$.

Example 9.1

Simple harmonic motion (SHM) with small damping is described, for example, by

$$\ddot{x} + x = -2\epsilon\dot{x}, \text{ with } x(0) = 1 \text{ and } \dot{x}(0) = 0 .$$

Apply Poincaré's method to this equation.

[The reason for considering SHM here is that we know the exact solution is

$$x = e^{-\epsilon t}\left\{\cos\sqrt{1-\epsilon^2}\,t + \frac{\epsilon}{\sqrt{1-\epsilon^2}}\sin\sqrt{1-\epsilon^2}\,t\right\} ,$$

and that the $\epsilon = 0$ solution is $x = \cos t$. The comparison between exact and approximate solution should prove useful.]

Let $x = x_0 + \epsilon x_1 + \ldots$ with $x_0(0) = 1,\ \dot{x}_0(0) = 0$

$$x_n(0) = \dot{x}_n(0) = 0, n \geqslant 1 \ ,$$

(i.e. we put all the initial conditions onto x_0 and let everything else have zero initial conditions).

Thus $(\ddot{x}_0 + \epsilon\ddot{x}_1 + \ldots) + (x_0 + \epsilon x_1 \ldots)$

$$= -2\epsilon(\dot{x}_0 + \epsilon x_1 + \ldots) \ ,$$

$$\underline{|\epsilon^\circ} : \quad \ddot{x}_0 + x_0 = 0 \ ,$$

$$\underline{|\epsilon} : \quad \ddot{x}_1 + x_1 = -2\dot{x}_0 \ ,$$

and so on.

From the first equation, $x_0 = \cos t$.

From the second equation, $\ddot{x}_1 + x_1 = 2 \sin t$ which has the solution

$$x_1 = \sin t - t \cos t, \text{ and so on.}$$

Thus $x = \cos t + \epsilon (\sin t - t \cos t) + \ldots$.

Graphing for various values of ϵ again shows poor agreement when either ϵ or t gets large, as shown in Fig. 9.3.

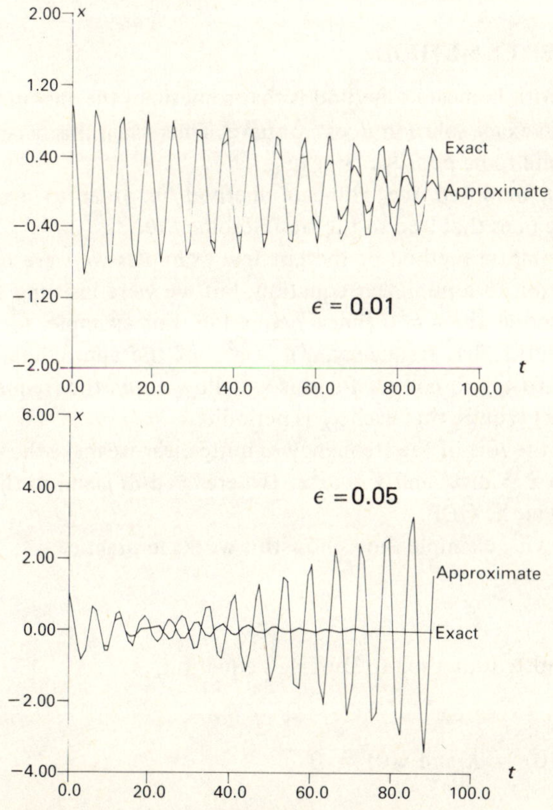

Fig. 9.3 – Solution of $\ddot{x} + x = -2\epsilon\dot{x}$, with $x(0) = 1$ and $\dot{x}(0) = 0$, for $\epsilon = 0.01, 0.05$.

Problems

Use Poincaré's method for the following equations.

1. $\ddot{x}+x = \epsilon x^3$ (Duffing's equation)

   with $x(0) = 1$ and $\dot{x}(0) = 0$.

2. $\ddot{x}+x = \epsilon(1-x^2)\dot{x}$ (Van der Pol's equation)

   with $x(0) = 1$ and $\dot{x}(0) = 0$.

[You may find the following identities useful:

$$\cos^3 t = \tfrac{1}{4}(3\cos t + \cos 3t),$$

$$\sin^3 t = \tfrac{1}{4}(3\sin t - \sin 3t) \text{ .]}$$

Answers to problems

1. $x = \cos t + \frac{\epsilon}{32}(12t \sin t + \cos t - \cos 3t) + \ldots$

2. $x = \cos t + \frac{\epsilon}{32}(12t \cos t - 9\sin t - \sin 3t) + \ldots$

9.3 LINDSTEDT'S METHOD

The trouble with Poincaré's method is that sometimes the answer contains terms in t which the exact solution does not have. Thus when this occurs the approximation is bound to be poor for large t.

Lindstedt modified the Poincare method in order to avoid SECULAR TERMS – the ones that lead to the troublesome t terms.

In the Poincaré method of the last few examples we were trying to find a periodic solution to a nonlinear equation, but we were insisting that it still had the same period as the $\epsilon = 0$ (linear) equation. For example, for damped SHM the exact solution has frequency $\sqrt{1-\epsilon^2}$ but the approximate solution has frequency 1. To try to correct for this we allow x *and* the frequency ω to vary with ϵ, and just require that each x_n is periodic.

To make the rôle of the frequency ω quite clear we make the transformation $\tau = \omega t$. Then $\dot{x} = \omega x'$ and $\ddot{x} = \omega^2 x''$ (where $' \equiv d/d\tau$), and so the parameter ω is now clearly in the ODE.

The following example shows how this works in practice.

Example 9.2

Apply the Lindstedt method to Duffing's equation,

$$\ddot{x} + x = \epsilon x^3 \quad ,$$

with $$x(0) = K \text{ and } \dot{x}(0) = 0 \text{ .}$$

Setting $\tau = \omega t$ gives

$$\omega^2 x'' + x = \epsilon x^3, \text{ with } x(0) = K \text{ and } x'(0) = 0 \ ,$$

and the frequency ω is now clearly visible.
Now we assume that

$$x = x_0 + \epsilon x_1 + \ldots$$

and $$\omega = 1 + \epsilon\omega_1 + \ldots ,$$

where $\omega_0 = 1$ since the $\epsilon = 0$ solution (i.e. SHM) has frequency 1.
The associated initial conditions are

$$x_0(0) = K, x_0'(0) = 0 \ ,$$

and $$x_n(0) = x_n'(0) = 0, \qquad n \geqslant 1 \ .$$

Substituting into the Duffing equation gives

$$(1 + \epsilon\omega_1 + \ldots)^2 (x_0'' + \epsilon x_1'' + \ldots)$$
$$+ (x_0 + \epsilon x_1 + \ldots) = \epsilon(x_0 + \epsilon x_1 + \ldots)^3 \ ,$$

and collecting like terms yields the following:

$$\underline{|\epsilon^\circ} : \ x_0'' + x_0 = 0$$

$$\underline{|\epsilon} : \ x_1'' + x_1 = x_0^3 - 2\omega_1 x_0''$$

$$\underline{|\epsilon^2} : \ x_2'' + x_2 = 3x_0^2 x_1 - 2\omega_1 x_1'' - (\omega_1^2 + 2\omega_2)x_0''$$

From the first equation, $x_0 = K \cos \tau$.
From the second equation,

$$x_1'' + x_1 = K^3 \cos^3 \tau + 2\omega_1 K \cos \tau$$

$$= \left(\frac{3K^3}{4} + 2\omega_1 K\right) \cos \tau + \frac{K^3}{4} \cos 3\tau \ .$$

The first term on the right-hand side is called the SECULAR TERM, and was causing all the problems above. To avoid it we set

$$\frac{3K^3}{4} + 2\omega_1 K = 0 \ ,$$

i.e. $$\omega_1 = -\frac{3K^2}{8} \ .$$

Then the second equation becomes $x_1'' + x_1 = (K^3/4) \cos 3\tau$, which has solution $x_1 = (K^3/32)(\cos\tau - \cos 3\tau)$.

From the third equation we have,

$$x_2'' + x_2 = \cos\tau\,(21K^5/128 + 2\omega_2 K) + \text{NST}\dagger,$$

where we have used the identities

$$\cos^2\tau = \tfrac{1}{2}(1 + \cos 2\tau)$$

$$\cos 2\tau \cdot \cos 3\tau = \tfrac{1}{2}(\cos\tau + \cos 5\tau)$$

and done a little algebra.
Again, to avoid secular terms we choose $\omega_2 = -21K^4/256$.
The third equation then becomes $x_2'' + x_2 = \text{NST}$, from which we can find

$$x_2 = \frac{K^5}{1024}(23\cos\tau - 24\cos 3\tau + \cos 5\tau)$$

and continue the procedure to whatever accuracy is required.

Thus

$$x(t) = K\cos\omega t + \frac{\epsilon K^3}{32}(\cos\omega t - \cos 3\omega t) + \ldots$$

where $$\omega = 1 - \frac{3\epsilon K^2}{8} + \ldots .$$

Note

It is important to notice that although this procedure can give very accurate answers (by deriving many terms in the asymptotic expansion), it does not give us *all* the solutions to the ODE. Since we assumed in the derivation that each term $x_n(t)$ was periodic, the Lindstedt method will only generate the *periodic* solutions, and will give us no information about how quickly the system tends to these solutions. Methods discussed in Chapters 10–12 *do* give us this further information – but at the expense of the simplicity of Lindstedt's method.

Problems

1. The equation

$$\ddot{x} + \frac{g}{l}\sin x = 0 ,$$

with $x(0) = K$ and $\dot{x}(0) = 0$, may be used to model the motion of a pendulum of length l swinging at a large angle x under gravity g.

(i) Set $T = \sqrt{g/l}\,t$ and substitute this into the above equation. This will eliminate the awkward g/l term.

† NST stands for nonsecular terms – those which cause no problems.

(ii) Now replace $\sin x$ by its Maclaurin series

$$\sin x = x - x^3/3! + \ldots$$

and compare the resulting equation with that in Example 9.2.
(iii) Hence solve the large-angle pendulum equation.

2. Apply Lindstedt's method to find the periodic solution of Van der Pol's equation

$$\ddot{x} + x = \epsilon(1 - x^2)\dot{x}\ , \quad \text{with}\ \dot{x}(0) = 0\ .$$

[Hint: We know that the Van der Pol equation has a limit cycle. Thus the periodic solution is unique and we must not specify $x(0)$. In fact, if you initially leave $x(0)$ arbitrary you will find that you need to choose $x_0(0) = 2$ in order to avoid secular terms.]

3. Apply Lindstedt's method to the differential equation

$$\ddot{x} + x = \epsilon x^2 \quad \text{with}\ x(0) = K \ \text{and}\ \dot{x}(0) = 0\ .$$

Hence show that the centre of the oscillation is at

$$x = \frac{\epsilon K^2}{6}(3 - 2\epsilon K), \text{ approximately}\ .$$

4. Apply Lindstedt's method to the differential equation

$$\ddot{x} + x = 1 + \epsilon x^2 \quad \text{with}\ x(0) = \dot{x}(0) = 0\ .$$

Hence show that the centre of oscillation is at

$$x = 1 + \frac{3\epsilon}{2} + \frac{13\epsilon^2}{3}, \text{ approximately}\ .$$

5. For the differential equation $\ddot{x} + x = \epsilon(x^2 + x^3/5)$ with $x(0) = K$ and $\dot{x}(0) = 0$, find the dependence of ω on K when $\epsilon = 0.1$.

Answers to problems

1. Setting $\epsilon = \frac{1}{6}$ gives

$$x = K\cos\omega T + \frac{K^3}{192}(\cos\omega T - \cos 3\omega T) + \ldots$$

with $$\omega = 1 - \frac{K^2}{16} + \ldots\ .$$

2. $$x = 2\cos\omega t + \frac{\epsilon}{4}(3\sin\omega t - \sin 3\omega t) - \frac{\epsilon^2}{96}(13\cos\omega t - 18\cos 3\omega t + 5\cos 5\omega t) + \ldots,$$

with $$\omega = 1 - \frac{\epsilon^2}{16} + \ldots.$$

3. $$x = K\cos\omega t + \frac{\epsilon K^2}{6}(3 - 2\cos\omega t - \cos 2\omega t) + \frac{\epsilon^2 K^3}{144}(-48 + 29\cos\omega t + 16\cos 2\omega t + 3\cos 3\omega t) + \ldots$$

with $$\omega = 1 - \frac{5\epsilon^2 K^2}{12} + \ldots.$$

4. $$x = 1 - \cos\omega t + \frac{\epsilon}{6}(9 - 8\cos\omega t - \cos 2\omega t) + \frac{\epsilon^2}{144}(624 - 509\cos\omega t - 112\cos 2\omega t - 3\cos 3\omega t) + \ldots$$

with $$\omega = 1 - \epsilon - \frac{23\epsilon^2}{12} + \ldots.$$

5. $$\omega = 1 - \frac{3K^2}{400}.$$

9.4 FORCED OSCILLATIONS

By applying exactly the same method, we can extend the ideas of section 9.3 to forced oscillations. Example 9.3 illustrates the technique.

Example 9.3

Consider Duffing's equation with soft resonant forcing

$$\ddot{x} + x = -\epsilon x^3 + \epsilon F\cos\omega t, \quad \omega \approx 1.$$

To simplify the algebra we restrict attention to the case $\dot{x}(0) = 0$.

Setting $\tau = \omega t$ as before and expanding via

$$x = x_0 + \epsilon x_1 + \ldots$$

and $$\omega = 1 + \epsilon\omega_1 + \ldots$$

with $x_n'(0) = 0$ gives the following.

$$\underline{|\epsilon^\circ} : \quad x_0'' + x_0 = 0$$

$$\underline{|\epsilon} : \quad x_1'' + x_1 = -2\omega_1 x_0'' - x_0^3 + F\cos\tau$$

$$\underline{|\epsilon^2} : \quad x_2'' + x_2 = -2\omega_1 x_1'' - 2\omega_2 x_0'' - \omega_1^2 x_0'' - 3x_0^2 x_1$$

From the first equation, $x_0 = A\cos\tau$.
From the second equation, $x_1'' + x_1 = (2\omega_1 A - 3A^3/4 + F)\cos\tau + \text{NST}$ and so we choose $\omega_1 = \frac{1}{2}(3A^2/4 - F/A)$ to eliminate the secular term.
The second equation can then be solved to give

$$x_1 = -\frac{A^3}{32}(\cos\tau - \cos 3\tau) \ .$$

Thus $$x = A\cos\omega t + \frac{\epsilon A^3}{32}(-\cos\omega t + \cos 3\omega t) + \ldots$$

with $$\omega = 1 + \frac{\epsilon}{2}\left(\frac{3A^2}{4} - \frac{F}{A}\right) + \ldots .$$

The expression above for ω gives the frequency as a function of amplitude A for fixed forcing ϵF. This is just the familiar AMPLITUDE RESPONSE CURVE, which shows how the output of a system varies with the frequency of the input.

For the linear case (no cubic term) we get $\omega = 1 - \dfrac{\epsilon F}{2A}$ which, for fixed drive ϵF, gives the familiar response curve shown in Fig. 9.4. The unbounded response at resonance (because of the lack of damping) is clearly visible.

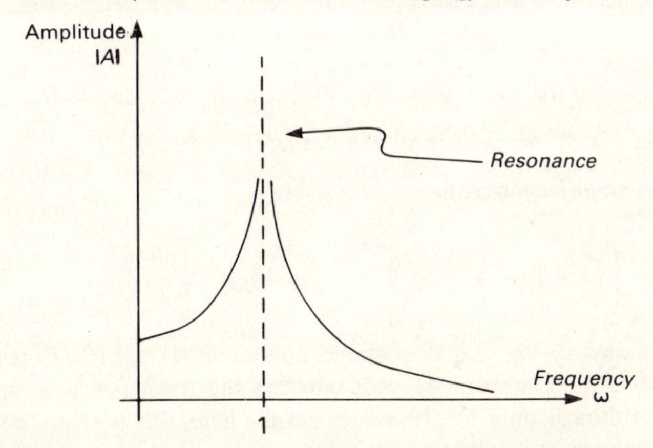

Fig. 9.4 – Amplitude response curve for forced simple harmonic motion.

For the nonlinear case, for ϵF fixed, the solution depends on the sign of ϵ. For $\epsilon > 0$ we have what is known as hard spring behaviour, and for $\epsilon < 0$ soft spring behaviour. These are shown in Fig. 9.5, where it is clear that the effect of the nonlinearity is to 'bend' the response curve obtained in the linear case. This is important since it means that, even with no damping, an infinite response is avoided when $\omega = 1$.

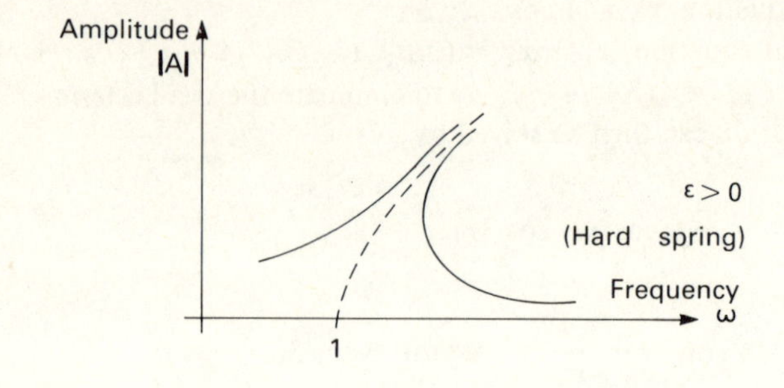

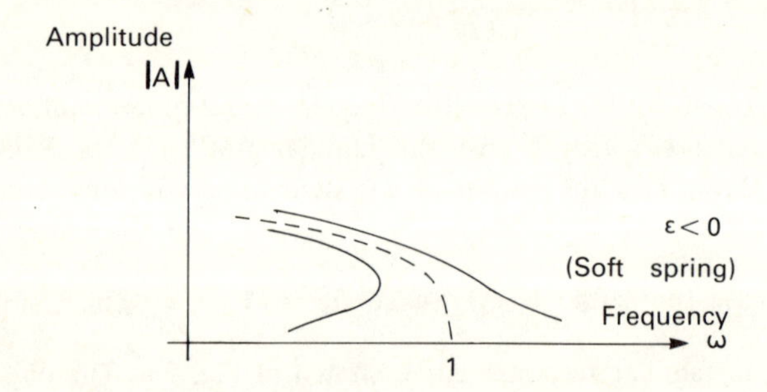

Fig. 9.5 – Amplitude response curve for Duffing's equation with soft resonant forcing.

If we repeated the whole procedure for the damped forced Duffing equation

$$\ddot{x} + x = \epsilon\{-2\delta\dot{x} - x^3 + F\cos\omega t\}$$

we would *eventually* deduce that

$$\left[\left(1 + \tfrac{3}{4}\epsilon A^2\right) - \omega^2\right]^2 + [2\epsilon\delta]^2 = \left[\frac{\epsilon F}{A}\right]^2 .$$

The graph shown in Fig. 9.6 now shows how jump HYSTERESIS is possible, where the amplitude history depends whether the frequency is increasing or decreasing. Although only mentioned in passing here, this is a very commonly observed phenomenon in vibrating systems.

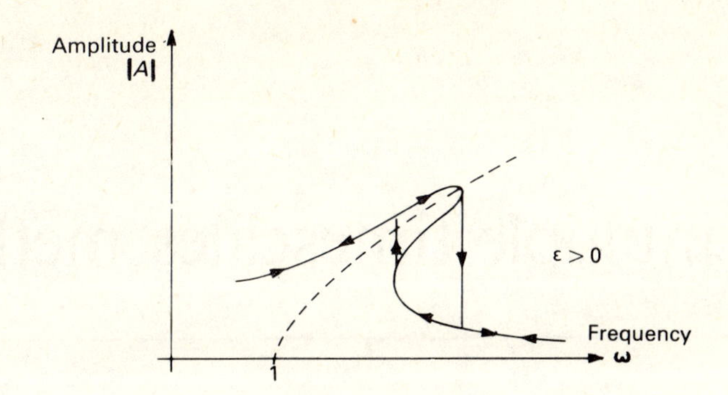

Fig. 9.6 – Amplitude response curve showing hysteresis.

Problem

Apply the above technique to the equation

$$\ddot{x} + x = \epsilon\{(1-x^2)\dot{x} + F\cos\omega t\}\ ,$$

the forced Van der Pol equation.

10

The multiple timescales method

10.1 INTRODUCTION

We mentioned earlier that the Lindstedt method will *only* generate the periodic solutions to a differential equation. Often this is enough, but sometimes it is necessary to discover how quickly the system tends to these solutions. The multiple timescales method of Chapter 10 and the method of Krylov and Bogoliubov in Chapter 11 supply this extra information.

10.2 THE METHOD

To introduce the essential ideas of the multiple timescales method, we will consider a situation we understand well: damped simple harmonic motion as modelled by

$$\ddot{x} + x = -2\epsilon\dot{x}$$

A Poincaré expansion, as above, sets

$$x = x_0 + \epsilon x_1 + \dots$$

and obtains the following results.

$$\lfloor \epsilon^0 : \quad \ddot{x}_0 + x_0 = 0$$

$$\lfloor \epsilon \ : \quad \ddot{x}_1 + x_1 = -2\dot{x}_0$$

$$\lfloor \epsilon^2 : \quad \ddot{x}_2 + x_2 = -2\dot{x}_1$$

and so on.

The first equation yields

$$x_0 = A\cos(t + \phi) \ ,$$

where A and ϕ are arbitrary constants; a solution of the second equation is

$$x_1 = -At\cos(t + \phi) \ ,$$

where we have chosen to retain only the particular integral in order to simplify the algebra. The third equation may then be solved to give

$$x_2 = \tfrac{1}{2}At^2 \cos(t+\phi) + \tfrac{1}{2}At \sin(t+\phi) \ .$$

Thus

$$x = A\cos(t+\phi) - \epsilon At\cos(t+\phi) + \tfrac{1}{2}\epsilon^2 A\,[t^2\cos(t+\phi) + t\sin(t+\phi)] + \ldots \ .$$

We note the appearance, once again, of the troublesome secular terms: x will be a poor approximation to the true solution when t is $O(\epsilon^{-1})$, since then ϵx_1 will no longer be smaller than x_0.

To see this more clearly we look at the exact solution

$$x = a\mathrm{e}^{-\epsilon t}\cos(\sqrt{1-\epsilon^2}\,t+\phi) \ .$$

Now

$$\mathrm{e}^{-\epsilon t} = 1 - \epsilon t + \frac{\epsilon^2 t^2}{2} + \ldots$$

and

$$\cos(\sqrt{1-\epsilon^2}\,t+\phi) = \cos(t+\phi) + \tfrac{1}{2}\epsilon^2 t\sin(t+\phi) + \ldots$$

from Maclaurin expansions in ascending powers of ϵ.

So we can write the expression above as

$$x = a\cos(t+\phi) - \epsilon\, at\cos(t+\phi) + \tfrac{1}{2}\epsilon^2 a[t^2\cos(t+\phi) + t\sin(t+\phi)] + \ldots \ .$$

Now a truncated $\mathrm{e}^{-\epsilon t}$ expression is only good if the combination ϵt is small, and so it seems as though ϵt is the important variable (and *not* t).
Similarly, $\epsilon^2 t$ is probably the important variable in the cosine expression.

We therefore take this hint and define new variables (i.e. multiple timescales) as follows.

$$T_0 = \epsilon^0 t = t$$

$$T_1 = \epsilon^1 t$$

$$T_2 = \epsilon^2 t$$

and so on.

Here T_{n+1} is always slower than T_n and, guided by the example above, we *assume* that the T_n's are independent.

To utilise these variables in the equation

$$\ddot{x} + x = -2\epsilon\dot{x} \ ,$$

we must use the chain rule for differentiating a function $F(T_0, T_1, T_2, \ldots)$ of many variables:

$$\frac{dF}{dt} = \frac{\partial F}{\partial T_0}\frac{dT_0}{dt} + \frac{\partial F}{\partial T_1}\frac{dT_1}{dt} + \frac{\partial F}{\partial T_2}\frac{dT_2}{dt} + \ldots$$

$$= \frac{\partial F}{\partial T_0} + \epsilon\frac{\partial F}{\partial T_1} + \epsilon^2\frac{\partial F}{\partial T_2} + \ldots .$$

We note that by using the D-operator

$$D \equiv \frac{d}{dt}$$

and defining similar operators

$$D_n \equiv \frac{\partial}{\partial T_n}$$

we may rewrite this last equation as

$$D = D_0 + \epsilon D_1 + \epsilon^2 D_2 + \ldots$$

in a similar form to our other expansions.

We may now substitute this expression into our equation

$$\ddot{x} + x = -2\epsilon\dot{x} ,$$

or the equivalent expression

$$D^2 x + x = -2\epsilon D x ,$$

to obtain

$$[(D_0 + \epsilon D_1 + \ldots)^2 + 1]\,[x_0 + \epsilon x_1 + \ldots]$$
$$= -2\epsilon[D_0 + \epsilon D_1 + \ldots]\,[x_0 + \epsilon x_1 + \ldots] .$$

As usual, we now extract like powers of ϵ.

$$\underline{|\epsilon^0} : \quad (D_0^2 + 1)\,x_0 = 0$$

$$\underline{|\epsilon^1} : \quad (D_0^2 + 1)\,x_1 = -2\,D_0 D_1 x_0 - 2\,D_0 x_0$$

and so on.

From the first equation we now have

$$x_0 = A\cos(T_0 + \phi) ,$$

where here A and ϕ are not constants but $A = A(T_1, T_2, \ldots)$ and $\phi = \phi(T_1, T_2, \ldots)$.

The second equation is then

$$(D_0^2 + 1)x_1 = 2\,D_1\,[A \sin(T_0 + \phi)] + 2A \sin(T_0 + \phi)$$
$$= 2\,[D_1 A + A]\,\sin(T_0 + \phi) + 2A\,D_1\phi \cos(T_0 + \phi)$$

using the product rule.
To remove secular terms we now set

$$D_1 A + A = 0 \text{ and } D_1\phi = 0\ ,$$

that is $$\frac{\partial A}{\partial T_1} + A = 0$$

and $$\frac{\partial \phi}{\partial T_1} = 0\ .$$

Thus $$A = a\,\mathrm{e}^{-T_1}$$

and $$\phi = c\ ,$$

where a and c are functions of all higher (and hence slower) timescales. In fact, a and c can be thought of as constant, at this timescale.

Substituting all this into the expressions above gives

$$x = x_0 + \epsilon x_1 + \ldots$$
$$= A \cos(T_0 + \phi) + \ldots$$
$$= a\,\mathrm{e}^{-T_1} \cos(T_0 + c) + \ldots\ .$$

So $x = a\,\mathrm{e}^{-\epsilon t} \cos(t + c)$, which agrees with the exact solution to $0(\epsilon)$.

Example 10.1
Duffing's equation is

$$\ddot{x} + x = -\epsilon x^3\ .$$

Substituting as above yields

$$(D_0^2 + 1 + 2\epsilon D_0 D_1 + \ \ldots)\,(x_0 + \epsilon x_1 + \ldots)$$
$$= -\epsilon(x_0 + \epsilon x_1 + \ldots)^3$$

$$\underline{|\epsilon^0} \ :\ (D_0^2 + 1)\,x_0 = 0$$

$$\underline{|\epsilon^1} \ :\ (D_0^2 + 1)\,x_1 = -2\,D_0 D_1\,x_0 - x_0^3$$

and so on.
From the first equation, we have $x_0 = A \cos(T_0 + \phi)$.

The second equation is then

$$\begin{aligned}
(D_0^2 + 1)x_1 &= -2D_0D_1[A\cos(T_0+\phi)] - A^3\cos^3(T_0+\phi)\\
&= 2D_1[A\sin(T_0+\phi)] - A^3\cos^3(T_0+\phi)\\
&= 2\,[A'\sin(T_0+\phi) + A\phi'\cos(T_0+\phi)]\\
&\quad -\frac{A^3}{4}[3\cos(T_0+\phi) + \cos 3(T_0+\phi)]\\
&= (2A')\sin(T_0+\phi) + \left(2A\,\phi' - \frac{3A^3}{4}\right)\cos(T_0+\phi) + \text{NST}\ .
\end{aligned}$$

Here NST, as previously, signifies nonsecular terms, and we have used a superscript $'$ to indicate $\partial/\partial T_1$.

To remove secular terms we now set

$$A' = 0 \quad \text{and} \quad \phi' = \frac{3A^2}{8}\ .$$

So A = constant = a and $\phi = (3a^2/8)\,T_1$ + constant, where by 'constant' we really mean a function of all the higher timescales.

Thus $$x = a\cos\left(T_0 + \frac{3a^2}{8}T_1 + \text{constant}\right) + \ldots$$

$$= a\cos\left(t + \frac{3\epsilon a^2}{8}t + \text{constant}\right) + \ldots\ .$$

We note that the result we derived earlier with the Lindstedt method may, with a slight change of variables, be written as

$$x = a\cos\omega t + \ldots$$

with $$\omega = 1 + \frac{3\epsilon a^2}{8} + \ldots\ ,$$

or as $$x = a\cos\left(t + \frac{3\epsilon a^2}{8}t + \ldots\right) + \ldots\ .$$

Thus the results are identical to $O(\epsilon)$.

Problems

Apply the method of multiple timescales to the following.

1. Van der Pol's equation $\ddot{x} + x = \epsilon(1 - x^2)\dot{x}$.
   [Hint: You should find that

$$A' = \frac{A}{2}\left[1 - \frac{A^2}{4}\right] .$$

   Multiplying by $2A$ this gives

$$Z' = Z(1 - Z/4), \text{ where } Z \equiv A^2 .$$

   Now use partial fractions to integrate.]
   Confirm that the limit cycle (i.e. the solution as $t \to \infty$) has amplitude 2.

2. $\ddot{x} + x = \epsilon(1 - x^2)\dot{x} + \epsilon F \cos \lambda t$, the Van der Pol equation with soft non-resonant forcing.
   [Hint: $\cos \lambda t = \cos \lambda T_0$]

3. $\ddot{x} + x = \epsilon(1 - x^2)\dot{x} + F \cos \lambda t$, the Van der Pol equation with hard non-resonant forcing.

   [Hint: Define $\eta = 1 - \dfrac{F^2}{2(1-\lambda^2)^2}$ and aim to get $A' = \dfrac{A}{2}\left[\eta - \dfrac{A^2}{4}\right]$.]

   Confirm that the steady-state solution ($t \to \infty$) depends on the sign of η.

4. For $\ddot{x} + 4x = \epsilon(3n - m\dot{x}^2)\dot{x}$, $m, n > 0$,
   show that the limit cycle has amplitude $\sqrt{n/m}$.

5. Consider the system described by the equation of motion

$$\ddot{x} + 3x + 6x^2 + 4x^3 = 0 .$$

   Use the method of multiple timescales to confirm that the solution near the origin is

$$x = \epsilon A \cos(\omega t + \phi) ,$$

   where $\omega^2 = 3 - 7\epsilon^2 A^2 + \ldots$

   and A, ϕ are constants.

Answers to problems;

1. $x = A \cos(t + \phi) + \ldots$

   with $A^2 = \dfrac{4}{1 + Be^{-\epsilon t}}$

   and ϕ, B constant (to this order of ϵ).

2. Answer as above. To this order, neither the amplitude nor the phase is affected.

3. $x = A \cos(t + \phi) + \dfrac{F}{1 - \lambda^2} \cos \lambda t + \ldots$

with $$A^2 = \frac{4\eta}{1 - Be^{-\epsilon\eta t}}$$

and ϕ, B constant (to this order of ϵ).

4. $x = A \cos(2t + \phi) + \ldots$

with $$A^2 = \frac{\eta}{m + B\,\mathrm{e}^{-3\eta\epsilon t}}$$

and ϕ, B constant (to this order of ϵ).

11

The method of Krylov and Bogoliubov

11.1 INTRODUCTION

Both the multiple timescales method and the method of Krylov and Bogoliubov give the complete time history of the solution (rather than merely the eventual periodic solution). If only first order solutions are required (i.e. those to $0(\epsilon)$), the method discussed in this chapter is rather more convenient.

11.2 THE METHOD

Although Krylov and Bogoliubov's method is fairly general, we will apply it only to equations of the form

$$\ddot{x} + \omega^2 x + \epsilon f(x, \dot{x}) = 0 \ , \tag{11.1}$$

where ϵ is small.

For the case $\epsilon = 0$ we may apply linear theory to obtain the solution

$$x = A \sin(\omega t + \phi) \ ,$$

where A and ϕ are arbitrary constants. Differentiating gives

$$\dot{x} = A\omega \cos(\omega t + \phi) \ .$$

Krylov and Bogoliubov suggested that, for small ϵ, the solution of (11.1) takes the form

$$x = A(t) \sin(\omega t + \phi(t)) \tag{11.2}$$

$$\dot{x} = A(t)\, \omega \cos(\omega t + \phi(t)) \tag{11.3}$$

where $A(t)$ and $\phi(t)$ are no longer constants but *slowly varying* functions of t.

Substituting $y = \dot{x}$ into (11.1) gives

$$\dot{y} = -\omega^2 x - \epsilon f(x, y) \tag{11.4}$$

and we look for a solution in the form

$$x = A(t) \sin(\omega t + \phi(t)) \tag{11.5}$$

$$y = A(t)\, \omega \cos(\omega t + \phi(t)) \ . \tag{11.6}$$

Substituting (11.5) and (11.6) into $y = \dot{x}$ gives

$$A\omega \cos(\omega t + \phi) = \dot{A} \sin(\omega t + \phi) + A(\omega + \dot{\phi}) \cos(\omega t + \phi) \ .$$

Thus

$$\dot{A} \sin(\omega t + \phi) + A\dot{\phi} \cos(\omega t + \phi) = 0 \ . \tag{11.7}$$

Substituting (11.5) and (11.6) into (11.4) gives

$$\dot{A}\omega \cos(\omega t + \phi) - A\omega(\omega + \dot{\phi}) \sin(\omega t + \phi) = -\omega^2 A \sin(\omega t + \phi) - \epsilon f(x, y)$$

Therefore

$$\dot{A} \cos(\omega t + \phi) - A\dot{\phi} \sin(\omega t + \phi) = -\frac{\epsilon}{\omega} f(A \sin(\omega t + \phi), A\omega \cos(\omega t + \phi)) \tag{11.8}$$

Solving (11.7) and (11.8) for $\dot{A}$ and $\dot{\phi}$, we get

$$\dot{A} = -\frac{\epsilon}{\omega} \cos(\omega t + \phi) \cdot f(A \sin(\omega t + \phi), A\omega \cos(\omega t + \phi)) \tag{11.9}$$

$$\dot{\phi} = \frac{\epsilon}{A\omega} \sin(\omega t + \phi) \cdot f(A \sin(\omega t + \phi), A\omega \cos(\omega t + \phi)) \ . \tag{11.10}$$

Note that $\dot{A}$ and $\dot{\phi}$ are both proportional to ϵ, confirming that A and ϕ are *slowly varying* functions of time when ϵ is small.

Note that, in terms of the assumptions contained in (11.2) and (11.3), equations 11.9 and 11.10 are *exact* representations of $\dot{A}$ and $\dot{\phi}$.

Krylov and Bogoliubov's approximation is to replace $\dot{A}$ and $\dot{\phi}$ in equations 11.9 and 11.10 by their *average values* over one period, $2\pi/\omega$. A is *regarded as a constant* in taking the average. This procedure (known as a method of averaging) leads to

$$\dot{A} = -\frac{\epsilon}{2\pi} \int_0^{2\pi/\omega} \cos(\omega t + \phi) \quad f(A \sin(\omega t + \phi), A\omega \cos(\omega t + \phi))\, dt \ ,$$

$$\dot{\phi} = \frac{\epsilon}{2\pi A} \int_0^{2\pi/\omega} \sin(\omega t + \phi) \quad f(A \sin(\omega t + \phi), A\omega \cos(\omega t + \phi))\, dt \ .$$

The substitution $\theta = \omega t + \phi$ gives the final results.

$$\dot{A} = -\frac{\epsilon}{2\pi\omega} \int_0^{2\pi} \cos\theta \quad f(A \sin\theta, A\omega \cos\theta)\, d\theta \tag{11.11}$$

$$\dot{\phi} = \frac{\epsilon}{2\pi A\omega} \int_0^{2\pi} \sin\theta \quad f(A \sin\theta, A\omega \cos\theta)\, d\theta \ . \tag{11.12}$$

The exact equations 11.9 and 11.10 are thus replaced by approximate equations 11.11 and 11.12. Once the integrals have been evaluated we have first order differential equations to solve for A and ϕ.

Evaluation of these integrals is simplified by making use of the following reduction formulae, which may be verified using integration by parts.

If we let $I_{m,n}$ denote $\int_0^{2\pi} \sin^m x \quad \cos^n x \, dx$,

then $$I_{m,n} = \frac{m-1}{m+n} I_{m-2,n} \tag{11.13}$$

and $$I_{m,n} = \frac{n-1}{m+n} I_{m,n-2} \ . \tag{11.14}$$

Equations 11.13 and 11.14 may be used until we reach either $I_{0,0} = 2\pi$ or $I_{0,1} = I_{1,0} = I_{1,1} = 0$.
Note that $I_{m,n}$ is only non-zero when m and n are *both* even.

Example 11.1
Apply the Krylov and Bogoliubov method to Van der Pol's equation

$$\ddot{x} + \omega^2 x - \epsilon(1-x^2)\dot{x} = 0 \ .$$

Comparing with the standard form

$$\ddot{x} + \omega^2 x + \epsilon f(x, \dot{x}) = 0$$

in (11.1) we have

$$f(x,\dot{x}) = -(1-x^2)\dot{x} \ .$$

Thus $f(A \sin\theta, A\omega\cos\theta) = -(1 - A^2 \sin^2\theta) A\omega \cos\theta$.
Substituting in (11.11) gives

$$\dot{A} = \frac{\epsilon}{2\pi\omega} \int_0^{2\pi} A\omega(1-A^2 \sin^2\theta) \cos^2\theta \, d\theta$$

$$= \frac{\epsilon A}{2\pi}(I_{0,2} - A^2 I_{2,2})$$

$$= \frac{\epsilon A}{2\pi}(\tfrac{1}{2} \cdot 2\pi - A^2 \cdot \tfrac{1}{4} \cdot \tfrac{1}{2} \cdot 2\pi) = \frac{\epsilon A}{8}(4 - A^2) \ .$$

This may be written as

$$\int \frac{dA}{A(2-A)(2+A)} = \frac{\epsilon}{8} \int dt \ ,$$

and using partial fractions gives

$$\int \left[\frac{1}{4A} + \frac{1}{8(2-A)} - \frac{1}{8(2+A)} \right] dA = \frac{\epsilon}{8} \int dt \ .$$

Performing the integrations gives

$$2 \ln A - \ln(2-A) - \ln(2+A) = \epsilon t + \ln C \ ,$$

or
$$\ln \left[\frac{A^2}{C(4-A^2)} \right] = \epsilon t \ .$$

Now suppose that $A = A_0$ when $t = 0$.

Then
$$C = \frac{A_0^2}{4-A_0^2} \ ,$$

and so
$$A^2 = \frac{\left(\dfrac{4A_0^2}{4-A_0^2} \right) e^{\epsilon t}}{1 + \left(\dfrac{A_0^2}{4-A_0^2} \right) e^{\epsilon t}} \ .$$

Finally,
$$A = \frac{2}{\left[\left(\dfrac{4}{A_0^2} - 1 \right) e^{-\epsilon t} + 1 \right]^{\frac{1}{2}}} \ .$$

Note that $A \to 2$ as $t \to \infty$, independent of the value of A_0. Thus the steady solution is a *limit cycle* of amplitude 2.

To gain information about the phase ϕ we return to equation 11.12.

$$\dot{\phi} = -\frac{\epsilon}{2\pi A\omega} \int_0^{2\pi} A\omega \, (1 - A^2 \sin^2\theta) \cos\theta \sin\theta \, d\theta \ ,$$

and so
$$\dot{\phi} = -\frac{\epsilon}{2\pi}(I_{1,1} - A^2 I_{3,1}) = 0 \ .$$

Thus $\phi = \phi_0 =$ constant, where ϕ_0 is the value of ϕ when $t = 0$.

The solution is given approximately by $x = A \sin(\omega t + \phi)$, and so

$$x \approx \frac{2 \sin(\omega t + \phi_0)}{\left[\left(\dfrac{4}{A_0^2} - 1\right) e^{-\epsilon t} + 1\right]^{\frac{1}{2}}} .$$

Example 11.2

Use the method of Krylov and Bogoliubov to find an approximate solution to the equation

$$\ddot{x} + x - \mu (x + 1)^2 = 0, \quad x(0) = -1, \quad \dot{x}(0) = 0 \ ,$$

given that μ is small.

Comparing with the standard form, we see that $\omega = 1$ and $f(x,\dot{x}) = -(x + 1)^2$.

We must convert the given initial conditions for x and $\dot{x}$ into conditions for A and ϕ. We are looking for a solution of the form

$$x = A \sin(t + \phi) \ ,$$

$$y = \dot{x} = A \cos(t + \phi) \ .$$

Thus when $t = 0, \quad -1 = A_0 \sin \phi_0$

and $\quad 0 = A_0 \cos \phi_0 \ .$

It is conventional to take $A_0 > 0$, and so

$$A_0 = 1, \quad \phi_0 = -\pi/2 \ .$$

The diagram in Fig. 11.1 shows that ϕ_0 is measured from the positive y-axis.

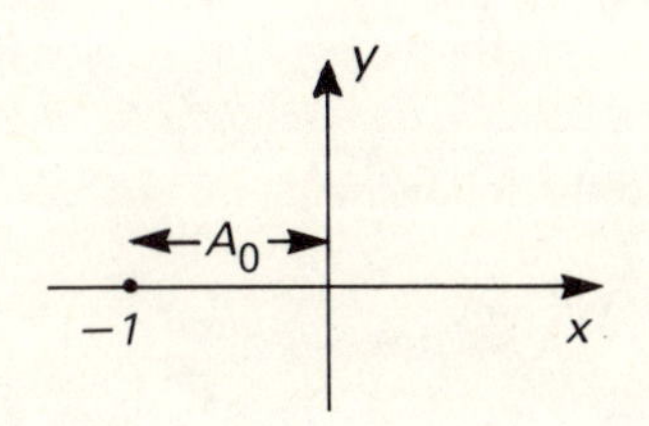

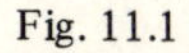
Fig. 11.1

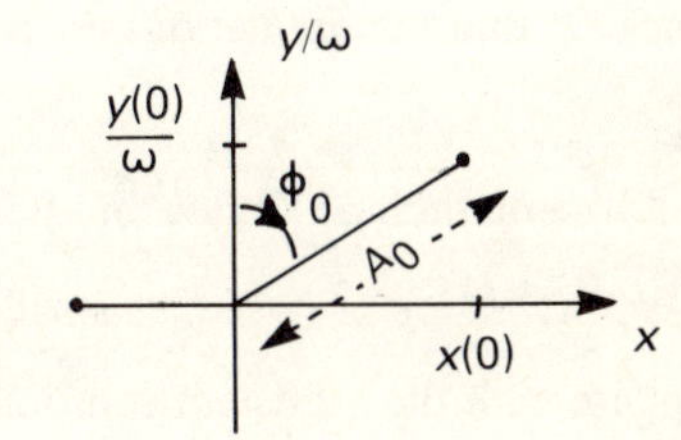

Fig. 11.2

In general the conversion from $x(0)$, $\dot{x}(0)$ to A_0, ϕ_0 can be seen diagrammatically in Fig. 11.2.

Equations (11.11) and (11.12) give

$$\dot{A} = \frac{\mu}{2\pi} \int_0^{2\pi} \cos\theta\,(A^2 \sin^2\theta + 2A\sin\theta + 1)\,d\theta \;,$$

$$\dot{\phi} = -\frac{\mu}{2\pi A} \int_0^{2\pi} \sin\theta\,(A^2 \sin^2\theta + 2A\sin\theta + 1)\,d\theta \;,$$

i.e. $$\dot{A} = \frac{\mu}{2\pi}[A^2 I_{2,1} + 2A I_{1,1} + I_{0,1}] = 0$$

and $$\dot{\phi} = -\frac{\mu}{2\pi A}[A^2 I_{3,0} + 2A I_{2,0} + I_{1,0}]$$

$$= -\frac{\mu}{\pi} \cdot \tfrac{1}{2} \cdot 2\pi = -\mu \;.$$

Integrating we get $A = A_0 = 1$,

$$\phi = -\mu t + \phi_0 = -\mu t - \pi/2 \;.$$

So $$x \approx \sin(t - \mu t - \pi/2) \;,$$

i.e. $$x \approx -\cos(1-\mu)t \;.$$

It is interesting to compare this solution with the solution given by Lindstedt's method,

i.e. $$x = -\cos\omega t + \mu\left(\tfrac{3}{2} - \tfrac{4}{3}\cos\omega t - \tfrac{1}{6}\cos 2\omega t\right) + \ldots$$

with $$\omega = 1 - \mu + \ldots \;.$$

The method of this section gives agreement on frequency to $0(\mu)$, but fails to pick up that the oscillations are not centred about $x = 0$.

Problems

Apply the method of Krylov and Bogoliubov to the following.

1. $\ddot{x} + \omega^2 x + \epsilon x^3 = 0$, ϵ small .

Compare with the Lindstedt solution.

2. $\ddot{x} + x + \epsilon\left(\tfrac{1}{3}\dot{x}^3 - \dot{x}\right) = 0$, $x(0) = 0$, $\dot{x}(0) = 1$.

For $\epsilon < 0$, what is the type and stability of the singular point $x = \dot{x} = 0$?

3. The *linear* equation $\ddot{x} + \epsilon\dot{x} + \omega^2 x = 0$. Compare with the exact solution.

4. $\ddot{x} - 2k\dot{x} + c\dot{x}^3 + \omega^2 x = 0$, k and c small.

Answers to problems

1. $x \approx A_0 \sin\left[\omega\left(1+\dfrac{3\epsilon A_0^2}{8\omega^2}\right)t+\phi_0\right]$.

2. $x \approx \dfrac{2\sin t}{\sqrt{3e^{-\epsilon t}+1}}$, stable focus .

3. $x \approx A_0 e^{-\epsilon t/2}\sin(\omega t+\phi_0)$.

4. $x \approx \dfrac{A_0 e^{kt}\sin(\omega t+\phi_0)}{\left[1+\dfrac{3c\omega^2A_0^2}{8k}(e^{2kt}-1)\right]^{\frac{1}{2}}}$.

11.3 STABILITY OF LIMIT CYCLES

The amplitudes of possible limit cycles are given by solutions of the equation

$$\dot{A} = 0, \quad \text{i.e. } A \text{ is constant} .$$

Now
$$\dot{A} = -\frac{\epsilon}{2\pi\omega}\int_0^{2\pi}\cos\theta \quad f(A\sin\theta, A\omega\cos\theta)\mathrm{d}\theta$$
$$= G(A), \quad \text{say} .$$

So the amplitudes of limit cycles are given by the solutions of $G(A) = 0$. The question arises whether or not the limit cycle is stable, i.e. if we made a slight disturbance from the limit cycle trajectory in the phase-plane, would the motion return to or diverge from the limit cycle.

Consider the expression for $\dot{A}$. Suppose that a solution of $G(A) = 0$ is $A = A_1$. A_1 is the amplitude of a limit cycle, and $G(A_1) = 0$.
Now make the disturbance $A = A_1 + \eta$ where η is small. For a *stable* limit cycle we require $\eta \to 0$ as $t \to \infty$.
Differentiating we obtain $\dot{A} = \dot{\eta}$.

Also
$$\dot{A} = G(A_1+\eta)$$
$$\approx G(A_1) + \eta G'(A_1)$$
$$= \eta G'(A_1)$$

since
$$G(A_1) = 0.$$

So
$$\dot{\eta} \approx \eta G'(A_1) .$$

Solving this equation gives

$$\eta \approx C e^{G'(A_1)t} ,$$

where C is an arbitrary constant.

So $\eta \to 0$ as $t \to \infty$, provided $G'(A_1) < 0$.

We now have a condition for stability:

if $G'(A_1) < 0$ there is a stable limit cycle at $A = A_1$,

if $G'(A_1) > 0$ there is an unstable limit cycle at $A = A_1$.

Example 11.3
Find the amplitude and stability of the limit cycle for Van der Pol's equation.

Van de Pol's equation is $\ddot{x} - \epsilon(1 - x^2)\dot{x} + \omega^2 x = 0$.

In Example 11.1 we obtained

$$\dot{A} = G(A) = \frac{\epsilon A}{8}(4 - A^2) .$$

Solutions of $G(A) = 0$ are $A = 0$, $A = 2$. Therefore we have a limit cycle of amplitude 0 (i.e. an equilibrium point) and a limit cycle of amplitude 2.

$$G'(A) = \frac{\epsilon}{8}(4 - 3A^2) .$$

For $\epsilon > 0$: $G'(2) < 0$, so $A = 2$ is stable limit cycle,

$G'(0) > 0$, so $A = 0$ (i.e. $x = \dot{x} = 0$) is an unstable equilibrium point.

For $\epsilon < 0$: $A = 2$ is an unstable limit cycle,

$A = 0$ is a stable equilibrium point.

Problems

1. Show that the solution of the nonlinear differential equation

$$\ddot{x} + 0.01(1 - x^2 + 0.05x^4)\dot{x} + x = 0$$

exhibits the following properties.

(i) A stable limit cycle of amplitude approximately 6.
(ii) An unstable limit cycle of amplitude approximately 2.1.
(iii) A stable point of equilibrium $x = \dot{x} = 0$.

2. Show that the system described by the differential equation

$$\ddot{x} + 4x + \epsilon(m\dot{x}^3 - 3n\dot{x}) = 0 \ ,$$

$$0 < \epsilon \ll 1, \quad m,n > 0 \ ,$$

possesses a stable limit cycle of amplitude $\sqrt{n/m}$.
Interpret this result when $n = 0$.

Answers to problems

1. $$\dot{A} = -\frac{\epsilon A}{320}(160 - 40A^2 + A^4) \ ,$$

$$\dot{\phi} = 0 \ .$$

2. $$\dot{A} = -\frac{3\epsilon A}{2}(A^2m - n) \ ,$$

$$\dot{\phi} = 0 \ .$$

When $n = 0$ we have $\dot{A} = -\dfrac{3\epsilon m}{2}A^3$,

with solution of the form $A^2 = \dfrac{1}{d + ct}$ $(c, d > 0)$,

so the origin is a stable equilibrium point.

12

Harmonic linearisation

12.1 INTRODUCTION

In the previous chapters we have presented several analytical methods. These are all quite complex, but have the feature that it is very easy to extend the solution to a higher order and hence generate solutions of arbitrary accuracy.

In this section we present the method of harmonic linearisation, which is also known as the describing function method in the control theory literature. This is probably the most powerful technique yet discussed, being able to handle not only the smooth (analytic) nonlinearities met so far but also the abrupt (non-analytic) nonlinearities met in many physical systems (see Part IV). Using this method it is extremely easy to obtain approximate solutions but much harder to improve them; we only do the former.

12.2 THE METHOD

We introduce the method via two examples.

Example 12.1

We consider the Duffing equation $\ddot{x} + x = \epsilon x^3, \dot{x}(0) = 0$, so that we can compare the new solution with previous knowledge.

Physical intuition leads us to expect a solution $x = A \cos \omega t$, which certainly fits the initial condition $\dot{x}(0) = 0$.

Then $x - \epsilon x^3 = A \cos \omega t - \epsilon A^3 \cos^3 \omega t$

$$= A \cos \omega t - \frac{\epsilon A^3}{4} [3 \cos \omega t + \cos 3\omega t]$$

$$= A \cos \omega t \left[1 - \frac{3\epsilon A^2}{4}\right] - \frac{\epsilon A^3}{4} \cos 3\omega t \ .$$

The third harmonic term here, the one containing $\cos 3\omega t$, as well as being small will be highly damped in any physical situation.

Thus $$x - \epsilon x^3 \approx A\cos\omega t \left[1 - \frac{3\epsilon A^2}{4}\right]$$

$$= x\left[1 - \frac{3\epsilon A^2}{4}\right],$$

and so we can write our equation as

$$\ddot{x} + \left[1 - \frac{3\epsilon A^2}{4}\right] x = 0 .$$

We have, however, assumed that the Duffing equation has solution $x = A\cos\omega t$, and so must have

$$\omega^2 = 1 - \frac{3\epsilon A^2}{4} .$$

Thus $$\omega \approx 1 - \frac{3\epsilon A^2}{8}$$

and so

$$x = A\cos\left(1 - \frac{3\epsilon A^2}{8}\right)t ,$$

in agreement with our previous results in Examples 9.2 and 10.1 and Problem 1 in section 11.2.

We note that the name 'harmonic linearisation' is a very accurate description of the procedure. The nonlinear term in the Duffing equation leads to harmonics being present in the solution, and we have linearised the equation by taking account only of the dominant terms. In Part IV we will encounter non-analytical nonlinearities, but the same ideas will apply – we will merely have to take a Fourier series of the nonlinear terms in order to identify their dominant parts.

Example 12.2

We apply harmonic linearisation to Rayleigh's equation

$$\ddot{x} + x = \epsilon(1 - \tfrac{1}{3}\dot{x}^2)\dot{x}, \quad \text{with } \dot{x}(0) = 0 .$$

Letting $x = A\cos\omega t$ we have $\dot{x} = -A\omega\sin\omega t$.

Then $x - \epsilon(1 - \frac{1}{3}\dot{x}^2)\dot{x}$

$$= A\cos\omega t + \epsilon A\omega\sin\omega t\ (1 - \tfrac{1}{3}A^2\omega^2\sin^2\omega t)$$

$$= A\cos\omega t + \epsilon A\omega\sin\omega t - \frac{\epsilon A^3\omega^3}{12}(3\sin\omega t - \sin 3\omega t)$$

So
$$x - \epsilon(1 - \tfrac{1}{3}\dot{x}^2)\dot{x}$$
$$\approx x - \epsilon\left(1 - \frac{A^2\omega^2}{4}\right)\dot{x}$$

Thus our original equation can be written as

$$\ddot{x} - \epsilon\left(1 - \frac{A^2\omega^2}{4}\right)\dot{x} + x = 0 .$$

For this to have an undamped periodic solution $x = A\cos\omega t$ we must have $\epsilon\,(1 - A^2\omega^2/4) = 0$ and $\omega = 1$. Thus $A = 2$ and our solution is $x = 2\cos t$ confirming the limit cycle of amplitude 2 for this equation. However, the method gives no information on the limit cycle stability, which depends on the sign of ϵ. Compare with Problem 2 in section 11.1.

Problems

1. Apply the method to Van de Pol's equation

$$\ddot{x} + x = \epsilon(1 - x^2)\dot{x}$$

with $\dot{x}(0) = 0$, and confirm that $x = 2\cos t$.

2. Apply the method to $\ddot{x} + x = \epsilon x^2$ with $\dot{x}(0) = 0$, and confirm that $\omega = 1$ with the centre of the oscillation at $x = \epsilon A^2/2$.

3. For $\ddot{x} + x = \epsilon\,(x^2 + x^3/5)$ with $\dot{x}(0) = 0$, confirm that when $\epsilon = 0.1$, $\omega = 1 - 3A^2/400$.

4. For the differential equation

$$\ddot{x} + x = (\tfrac{1}{2} - 3\dot{x}^2)\dot{x} - x^2 \quad \text{with}\ \dot{x}(0) = 0 ,$$

(a) determine the singular points,
(b) sketch trajectories near to the singular points,
(c) discuss the possibility of a limit cycle,
(d) if you think there is a limit cycle, use the method of harmonic linearisation to find a first order approximation.

Answers to problems

4. (a) Singular points are (0,0) and (−1,0).
   (b) Phase-plane methods give (0,0) as an unstable focus and (−1,0) as a saddle.
   (c) Phase plane diagram shows a limit cycle might exist, centred near the origin and with amplitude between 0 and 1.
   (d) $A = \dfrac{\sqrt{2}}{3} \approx 0.471$.

Part IV

Oscillations in certain practical situations

13

Self-excited oscillations

13.1 INTRODUCTION

The forced, damped SHM equation

$$m\ddot{x} + c\dot{x} + kx = F \cos \omega t$$

describes the response of a linear system to a periodic loading, and is easily studied using standard methods (see, for example, Dunning-Davies (1982)). The equation might, for instance, represent the response of a car's suspension system to a bumpy road; the RHS is known as the forcing function, and the periodic forcing function here comes from the periodic bumpiness of the road. Since the forcing function is periodic, it is not surprising that the solution to this equation (i.e. the motion of the car) is eventually also periodic.

In this chapter we are interested in similar equations representing oscillations, the difference being that although the solution is still periodic the forcing function this time is not. This may sound a little unusual, but we are surrounded by many everyday examples of this sort of thing – for instance, the chalk moving in one direction on the blackboard leads to a vibration and hence to a screech. Similar examples are the chattering of brake shoes in a car, the electric doorbell run from a DC source, the shimmying of the wheels on a supermarket trolley, the vibrations of a violin string caused by the bow, and the squeaking of an unoiled hinge. Since all of these lack a periodic forcing function (or drive), the term SELF-EXCITED OSCILLATIONS is used to describe them.

The next section discusses examples of self-excited oscillations, most of which can be described by linear differential equations with constant coefficients – the easy case! In the later sections we go on to consider systems with nonlinear characteristics.

13.2 INSTABILITY CAUSED BY FRICTION

Most of the examples mentioned in the introduction were of this type, one of the best known being the violin string being excited by a bow. The string is the vibrating system and the steady pull of the bow is the required non-alternating source of energy. The friction between the string and the bow has the characteristic of being greater for small slipping velocities (i.e. relative velocities of bow and string) than for large ones. This property of dry friction is completely opposite to that of viscous friction (where the force is proportional to the velocity), and is shown in Fig. 13.1.

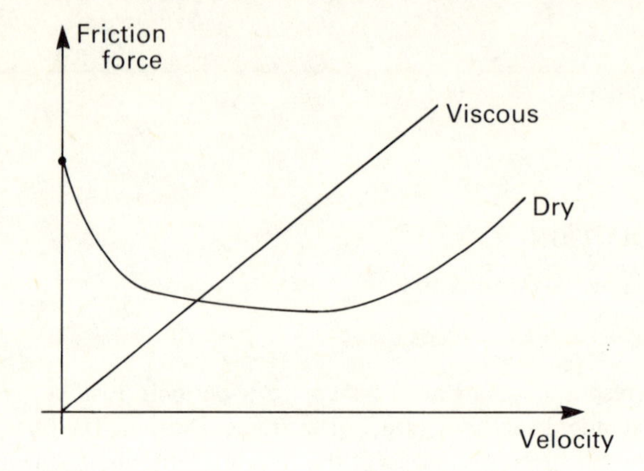

Fig. 13.1 – Behaviour of friction force with velocity.

Consider the bow moving at a constant speed over the vibrating string. While the string is moving in the direction of the bow, the slipping velocity is small and consequently the friction force great; during the receding motion of the string, the slipping velocity is large and the friction small. Note that the large friction force acts on the string in the direction of its motion, whereas the small friction force acts against the motion of the string. Thus we have a mechanism for feeding energy into the motion of the string, and so the vibration will build up.

13.3 GALLOPING OF TRANSMISSION LINES

HT electrical transmission lines have been observed, under certain weather conditions, to vibrate with great amplitudes and at very low frequencies; this is known as galloping. The line consists of a wire stretched between towers about 100 m apart. A span of the line will vibrate with one or two half-waves with an amplitude as great as 3 m in the centre at a rate of around 1 Hz.

Fig. 13.2 – Galloping of transmission lines.

When the wind blows against a circular cylinder (like the wire) it exerts a force in the same direction as itself – the drag force. Consider the transmission line during the downward stroke of its galloping motion moving at velocity v. If there is a sidewind of velocity V, the wire will experience a wind blowing at an angle $\alpha = \tan^{-1} (v/V)$ slightly from below and a drag force in the same direction, as shown in Fig. 13.3. Since the wire is moving downwards and the drag force has an upward component, the force exerted by the wind has a damping effect.

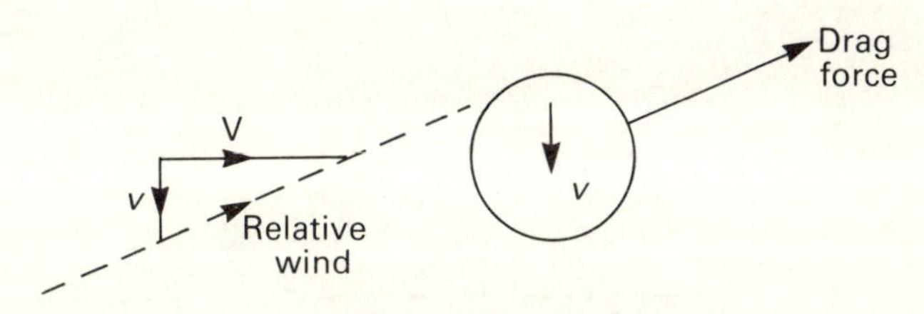

Fig. 13.3 – Forces on a circular cylinder.

In most cases where galloping has been observed, however, sleet has been found on the wire. Thus the wire is no longer circular in cross-section and so both lift and drag forces are induced. Taking components of these forces shows that the total upward force of the wind is $F = L \cos\alpha + D \sin\alpha$. We are not interested

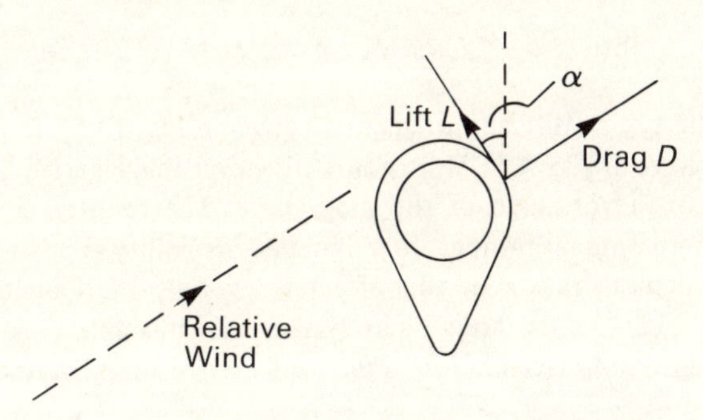

Fig. 13.4 – The effect of icing on the forces.

in the force F itself, since it is merely helping to carry the weight of the line, but rather how it varies with α. That is, we are interested in $\mathrm{d}F/\mathrm{d}\alpha$.

$$\frac{\mathrm{d}F}{\mathrm{d}\alpha} = \frac{\mathrm{d}L}{\mathrm{d}\alpha} \cos\alpha - L \sin\alpha + \frac{\mathrm{d}D}{\mathrm{d}\alpha} \sin\alpha + D \cos\alpha$$

$$= \sin\alpha \left[-L + \frac{\mathrm{d}D}{\mathrm{d}\alpha}\right] + \cos\alpha \left[\frac{\mathrm{d}L}{\mathrm{d}\alpha} + D\right]$$

$$\approx \frac{\mathrm{d}L}{\mathrm{d}\alpha} + D, \text{ since } \alpha \text{ is small} \; .$$

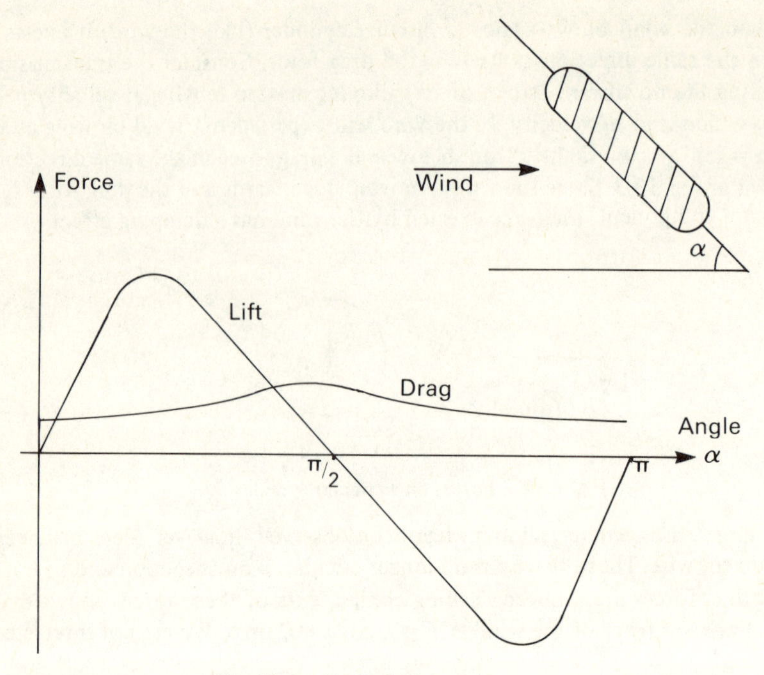

Fig. 13.5 – Typical lift and drag curves.

Examining the typical lift and drag curves given in Fig. 13.5 shows that it is quite possible to have $\mathrm{d}F/\mathrm{d}\alpha < 0$. This occurs whenever the negative slope of the lift is greater than the ordinate of the drag curve. The result is a force that encourages the alternating motion: again a dynamic instability is possible.

A very neat demonstration of this effect is given by the Lanchester tourbillion, shown in Fig. 13.6, where an aerodynamically unstable cross-section is pivoted in the middle and placed before a fan; self-excited rotation results.

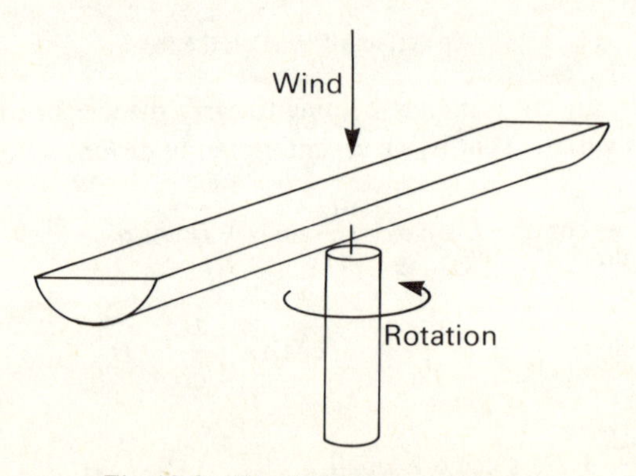

Fig. 13.6 – The Lanchester tourbillion.

13.4 VORTEX SHEDDING

When a fluid flows past a cylindrical obstacle, the wake behind the obstacle is no longer regular but can contain vortices as shown in Fig. 13.7. The vortices

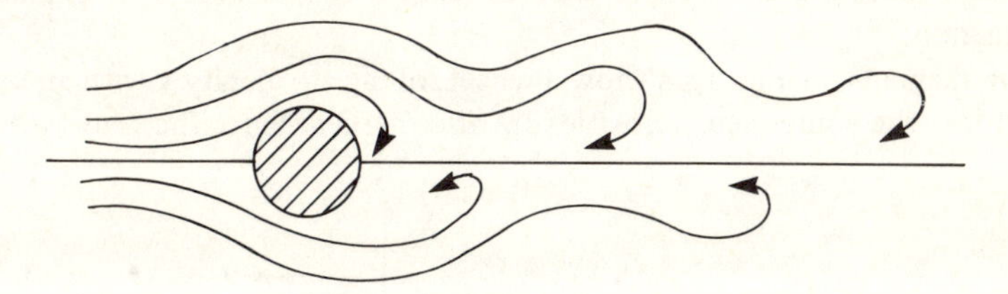

Fig. 13.7 – Vortex shedding from a circular cylinder.

are alternately clockwise and anticlockwise, being shed from alternate sides of the cylinder in a regular manner. They are thus associated with an alternating sideways force at the shedding frequency. Experiments with a cylinder show that an important parameter is the Strouhal number S given by $S = fD/V \approx 0.2$, where f is the shedding frequency, D the cylinder diameter and V the flow velocity. Thus once again a non-alternating drive has produced an alternating force.

Consider now a HT transmission line of 2 cm diameter in a 15 m/s wind – both common values. The formula above indicates that the shedding frequency will be about 150 Hz. The ensuing force will drive the line into resonance at a high harmonic (typically the 25th) and at low amplitude. Thus a simple damping device is sufficient to stop fatigue failures, and is usually incorporated into the design of the transmission line.

For a steel industrial smokestack, however, the problem is more severe. For a diameter of 5 m and windspeed of 15 m/s, the Strouhal frequency turns out to be about 0.6 Hz – very near the natural frequency of the structure. Thus great care has to be taken to minimise vortex shedding and strakes (or splitter plates) can often be seen on factory chimneys, which aim to mess up the shedding process and ensure that vortices up and down the cylinder are not all in phase.

This same vortex shedding mechanism was responsible for the Tacoma Narrows bridge disaster in the USA in 1940, where a Strouhal frequency near the natural frequency caused large amplitude oscillations and subsequent failure. Braun (1975) gives an interesting discussion.

13.5 THE SIMPLE CASTOR

A steered rolling wheel, whose point of contact with the ground lies behind the point of intersection of the steering axis and the ground, is usually called a castor. Examples range from simple supermarket trolleys (as shown in Fig. 13.8) to aircraft landing gear and motorcycle front wheels.

The rapid build up of high frequency oscillations in a castor, accompanied by a violent shudder in the main driving body, is called SHIMMYING – an undesirable nuisance for supermarket trolleys but a serious structural problem in aircraft landing gear. We will here develop a simple model to explain this phenomenon.

The diagrams in Fig. 13.8 show a wheel rolling at velocity **v** with an oscillating castor – the component velocities are also marked. Since the wheel is rolling,

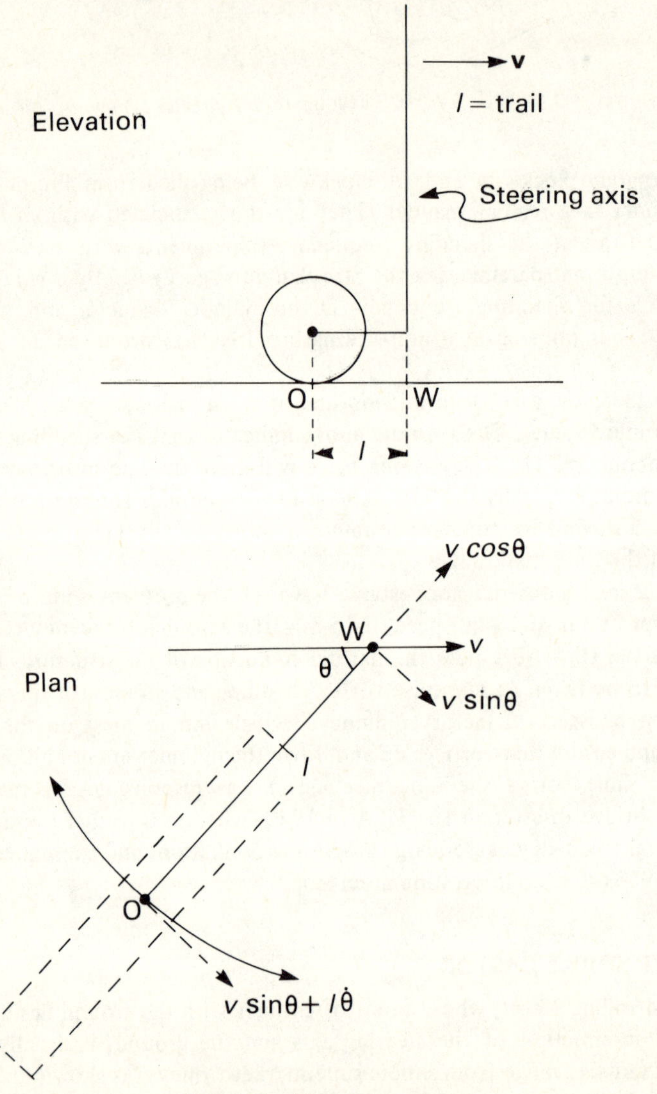

Fig. 13.8 – The simple castor.

we have no friction in the direction OW. In the direction perpendicular to this, however, the wheel must be slipping, and so there is a friction force μR sign $(v \sin\theta + l\dot{\theta})$. Here we have assumed Coulomb (or dry) friction and let the reaction at W be R. If we denote the moment of inertia about the vertical through W by I, Newton's second law gives

$$I\frac{d^2\theta}{dt^2} = -\mu R l \,\mathrm{sign}(v \sin\theta + l\dot{\theta}) \ .$$

We non-dimensionalise this equation by setting

$$\tau = \frac{vt}{l}, \ x = \frac{Iv^2\theta}{\mu R l^3} = B\theta \ .$$

We then have

$$\frac{d^2x}{d\tau^2} = -\,\mathrm{sign}\left[\frac{1}{B}\frac{dx}{d\tau} + \sin\frac{x}{B}\right] \quad \text{for } v > 0 \ .$$

In a typical application, $B \gg 1$. Thus, for small angles θ, x/B is small and so $\sin x/B \approx x/B$.
We then have $x'' = -\mathrm{sign}(x + x')$, since $B > 0$.

This is very similar to the example we met in section 7.2, with phase portrait as shown in Fig. 13.9. Thus the shimmy is stable and the oscillations will soon die out.

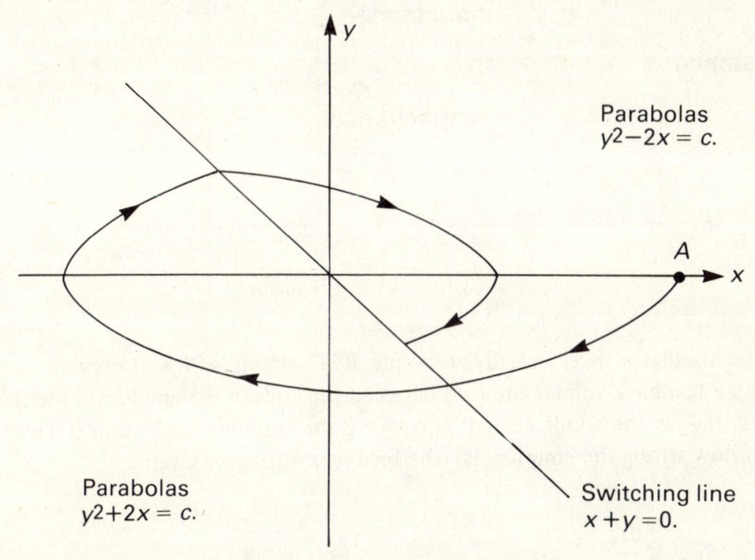

Fig. 13.9 – The phase-plane solution.

For moderate angles of shimmy, the switching line is

$$y + B \sin \frac{x}{B} = 0 \ .$$

This has a vastly different graph, depending on the size of B, and more care must be taken to ensure stability. Further details are given in Oke (1981).

13.6 THE VALVE OSCILLATOR

Although superseded by solid state devices for most applications, the valve oscillator shown in Fig. 13.10 is still regularly used for high-power devices. We here discuss its nonlinear characteristics.

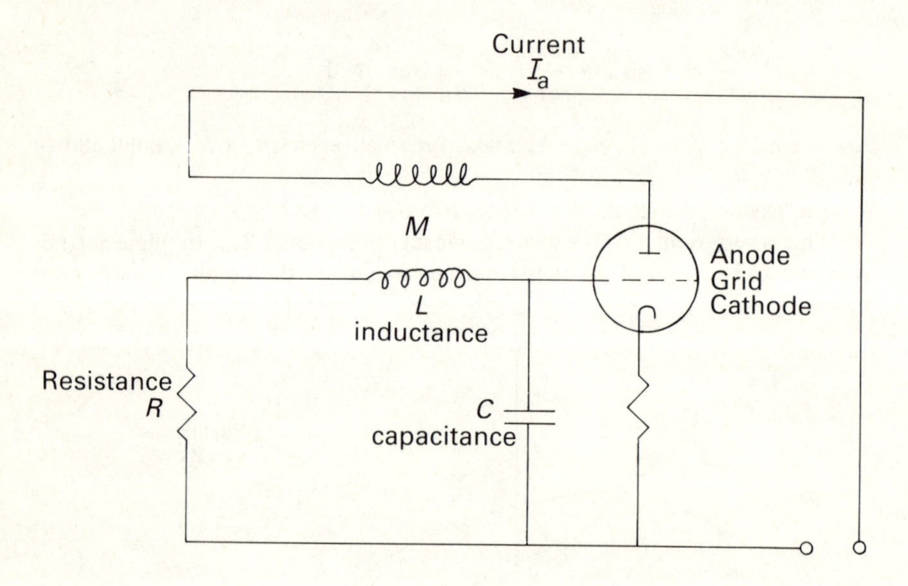

Fig. 13.10 – A valve oscillator.

The oscillator is essentially a simple RLC circuit, whose losses are made good by a feedback voltage induced between the coils in the anode and oscillator circuits. If I_a is the anode current and $M > 0$ the coupling coefficient (an indication of how strong the coupling is), the feedback voltage is given by

$$-M \frac{dI_a}{dt} \ .$$

Thus the voltage equation can be written as

$$L\ddot{Q} + R\dot{Q} + \frac{1}{C}Q - M\frac{dI_a}{dt} = 0 \ ,$$

as shown in Strain (1961).

The relationship between anode current I_a and grid voltage V_g is given by the valve characteristic in Fig. 13.11, where $(\bar{V}_g, \bar{I}_a)$ is the working point of the system. Measuring grid voltages from this working point (i.e. $x \equiv V_g - \bar{V}_g = Q/C$), we can use the fact that

$$\frac{dI_a}{dt} = \frac{dI_a}{dx} \cdot \frac{dx}{dt}$$

to rewrite the above equation as

$$LC\ddot{x} + RC\dot{x} + x - MS(x)\dot{x} = 0 \ .$$

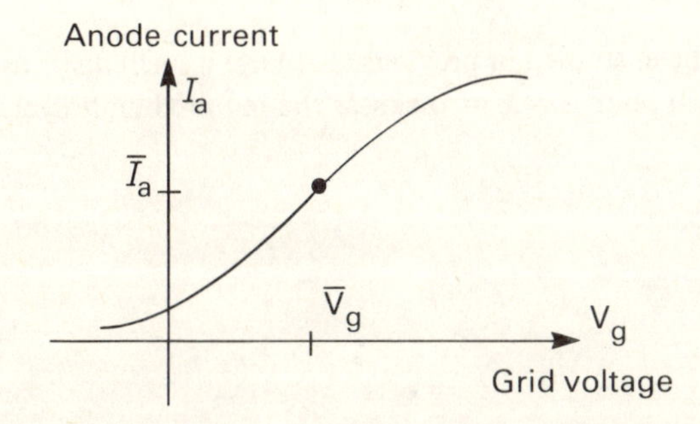

Fig. 13.11 – Typical valve characteristic.

Here $S(x) \equiv dI_a/dx$ is the slope of the valve characteristics and is shown in Fig. 13.12.

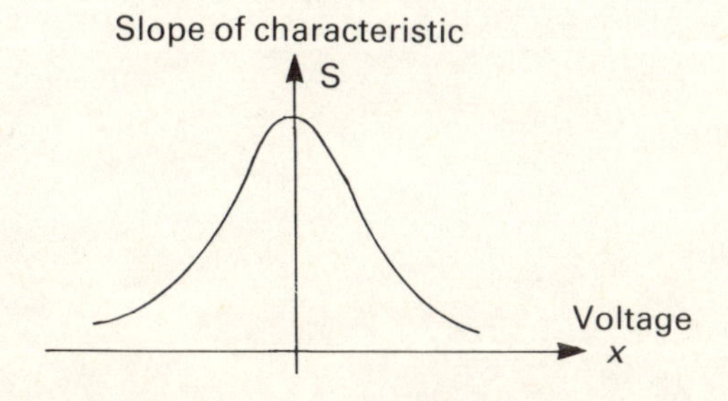

Fig. 13.12 – Slope of valve characteristic.

By putting $t = \tau\sqrt{LC}$ and $D = R\sqrt{C/L}$ the equation can be written in dimensionless form as

$$x'' - \left[\frac{MS(x)}{\sqrt{LC}} - D\right] x' + x = 0 .$$

The slope $S(x)$ here is, approximately, an even function (see Fig. 13.12), and so we may use a Taylor series to write

$$S(x) = S_0 + S_2 x^2 + S_4 x^4 + \ldots$$

and approximate the above equation by Van der Pol's equation

$$x'' - (\alpha + \beta x^2)x' + x = 0 ,$$

where

$$\alpha = \frac{MS_0}{\sqrt{LC}} - D \text{ and } \beta = \frac{MS_2}{\sqrt{LC}} < 0 .$$

This equation has been studied in previous chapters: a small disturbance from the unstable equilibrium point $x = x' = 0$ causes the required limit cycle oscillation.

14

Large nonlinearities

14.1 INTRODUCTION

We have seen that Van der Pol's equation

$$\ddot{x} - \epsilon(1-x^2)\dot{x} + x = 0, \ \epsilon \ small \ ,$$

exhibits a limit cycle oscillation of amplitude approximately 2. When ϵ is *large,* however, the behaviour is totally different: the motion becomes jerky, and the periodic solution is known as a relaxation (or Kipp) oscillation. A typical experiment to demonstrate this is discussed below.

14.2 KIPP OSCILLATOR EXPERIMENT

The hydraulic system shown in Fig. 14.1 is extremely simple, but will illustrate the main features of relaxation (or Kipp or sawtooth) oscillations. A storage tank is filled by a steadily flowing stream of water, until at a water height h_2 a syphon operates and causes the level to drop to a height h_1. Air entering the syphon then breaks off the discharge and the cycle repeats. Thus the graph of

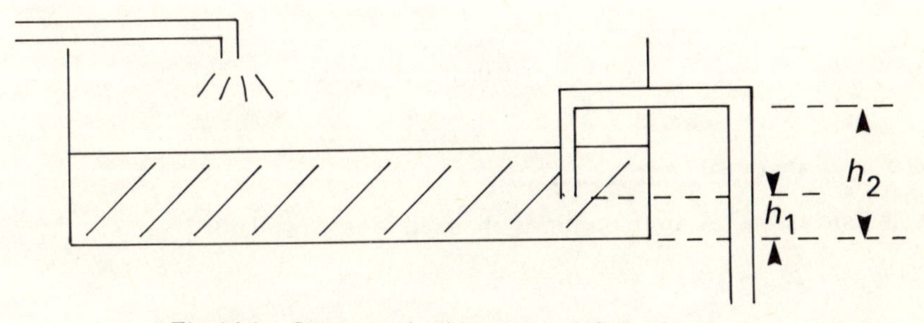

Fig. 14.1 – Storage tank with constant inflow of water.

water depth h against time looks rather like a sawtooth, with the period of oscillation being simply the sum of filling time T_F and emptying time T_E. This is shown in Fig. 14.2.

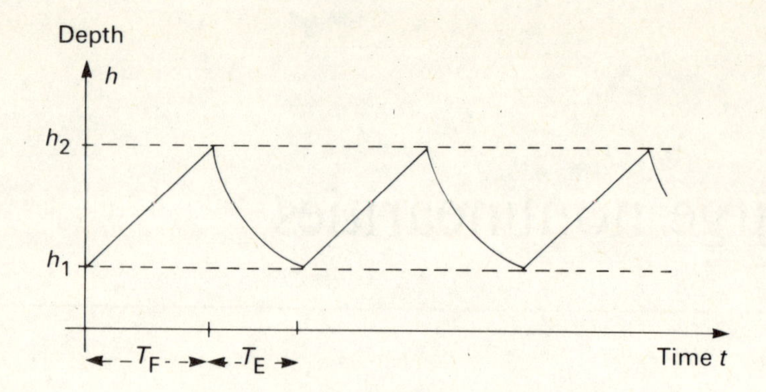

Fig. 14.2 – Typical relaxation oscillations.

This limit cycle behaviour is shown in the phase-plane diagram, see Fig. 14.3.

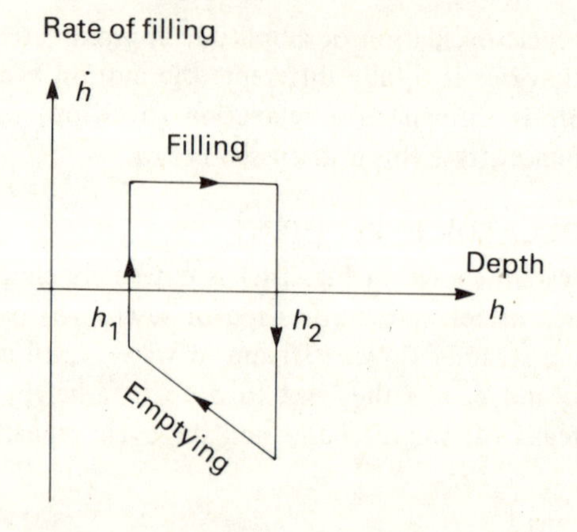

Fig. 14.3 – The limit cycle behaviour.

14.3 RELAXATION OSCILLATIONS

A similar type of sawtooth output is obtained for the oscillator

$$\ddot{x} + \epsilon F(\dot{x}) + x = 0$$

when the damping term F has a particular form and when ϵ is *large.*

This equation is, of course, in the correct form to apply Lienard's method where (knowing the form of $\epsilon F(y)$) we plot $x + \epsilon F(y) = 0$ in the phase plane and use a particular graphical construction (see Section 6.3) to build up the solution curves. However, in the present case we know that ϵ is large, and we can use this fact to short-cut the graphical method. To do this we note that the solution slope is

$$\frac{\mathrm{d}y}{\mathrm{d}x} = \frac{\mathrm{d}y/\mathrm{d}t}{\mathrm{d}x/\mathrm{d}t} = \frac{\dot{x}}{y} = \frac{-[x + \epsilon F(y)]}{y}$$

and $[x + \epsilon F(y)]$ is very large away from the curve $x + \epsilon F(y) = 0$. Thus the solution will consist approximately of vertical lines (slope $= \infty$) and horizontal lines (slope $= 0$).

Example 14.1

Consider the example where $F(y)$ is defined as in Fig. 14.4. To sketch $x + \epsilon F(y) = 0$, merely rotate this curve through $\pi/2$ anticlockwise and rescale the x axis by a factor ϵ. [An easy way to see this is to plot points corresponding to $x = 0, \pm 1, \pm \epsilon$, i.e. $F(y) = 0, \mp 1/\epsilon, \mp 1$.] We then have the graph shown in Fig. 14.5.

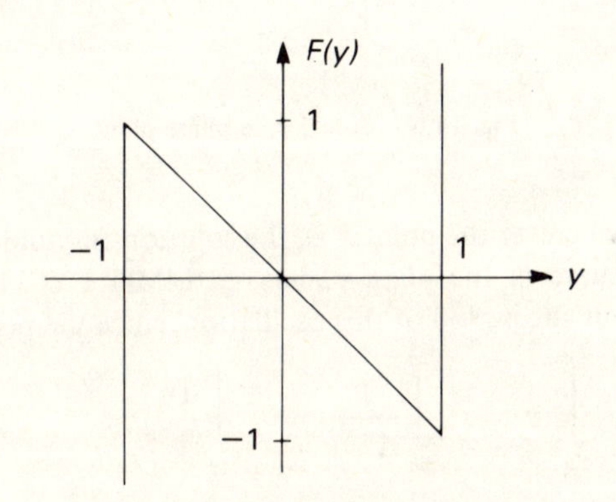

Fig. 14.4 – The nonlinearity $F(y)$.

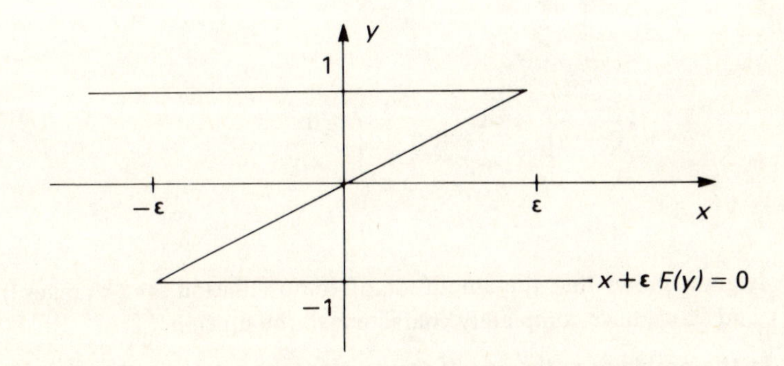

Fig. 14.5 – Sketch of $x + \epsilon F(y) = 0$ in the phase plane.

Starting at some arbitrary point P away from the curve $x + \epsilon F(y) = 0$, the solution will consist of vertical lines and horizontal lines as shown in Fig. 14.6.

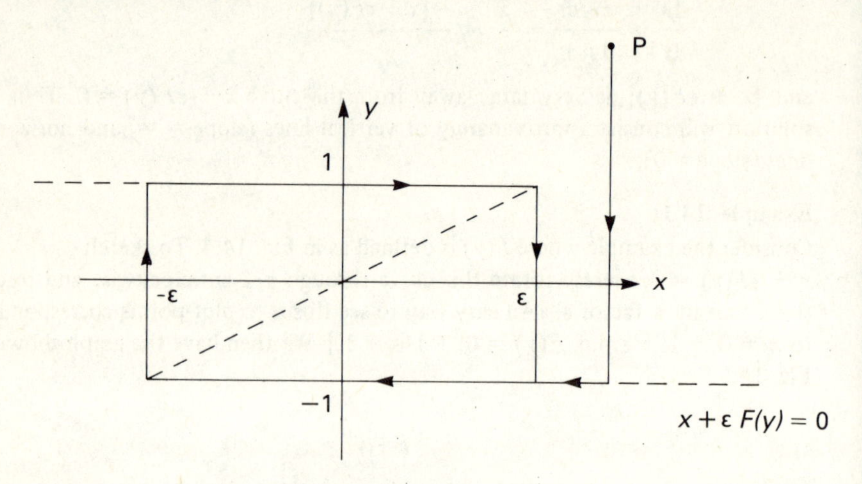

Fig. 14.6 – Solution in phase-plane.

Thus we see that wherever the point P is, the solution very quickly tends towards a unique closed curve in the phase-plane – a LIMIT CYCLE. We can, in fact, very easily work out the period of this oscillation, since the period will be

$$T = \int_{\text{cycle}} \mathrm{d}t = \int_{\text{cycle}} \frac{\mathrm{d}x}{\mathrm{d}x/\mathrm{d}t} = \int_{\text{cycle}} \frac{\mathrm{d}x}{y} .$$

In vertical portions of the solution curve we know that $\mathrm{d}x = 0$, and on horizontal portions we know the equation of the curve.

Thus
$$T = \int_{\text{cycle}} \frac{\mathrm{d}x}{y}$$

$$= 0 + \int_{\epsilon}^{-\epsilon} \frac{\mathrm{d}x}{-1} + 0 + \int_{-\epsilon}^{\epsilon} \frac{\mathrm{d}x}{+1}$$

$$= 4\epsilon .$$

The diagram shows that the amplitude of the oscillation is ϵ (x ranges from $-\epsilon$ to ϵ), and so we have completely characterised the motion.

In the problems at the end of this section, the situation at first sight appears more complex. In one of these problems we have the situation shown in Fig. 14.7.

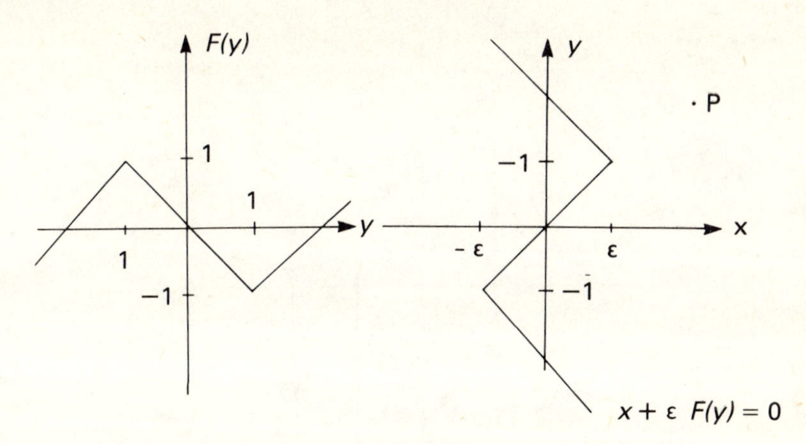

Fig. 14.7 – The nonlinearity $F(y)$.

What happens when the vertical portion of the solution curve from P breaks the slanting curve $x + \epsilon F(y) = 0$? Nothing special – the solution is still vertical portions away from the curve and horizontal portions near the curve. We thus get a sort of staircase until the solution becomes vertical, see Fig. 14.8. Since this

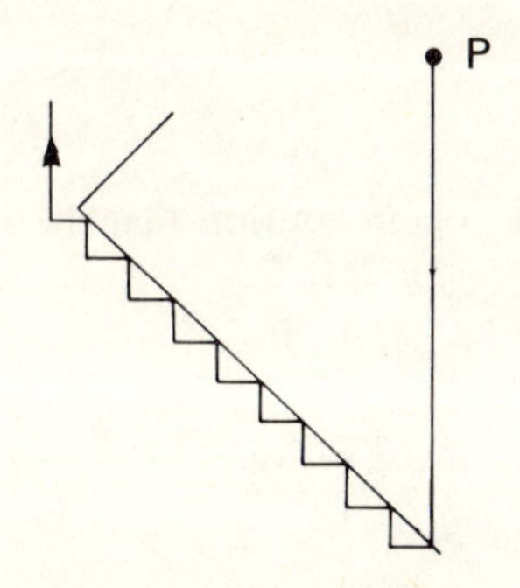

Fig. 14.8 – Enlargement of part of phase-plane.

staircase is extremely small, the effect is that the solution follows the curve itself. Thus the limit cycle in this case is simply as shown in Fig. 14.9. To evaluate the period in this case, we note again that $dx = 0$ on vertical portions and on sloping portions we know the equation of the curve and so can evaluate

$$\int \frac{dx}{y} \quad \text{or} \quad \int \frac{1}{y}\frac{dx}{dy}\,dy \;.$$

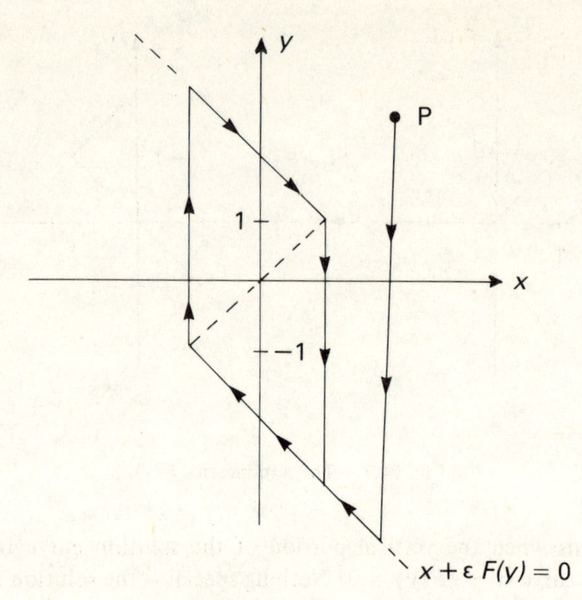

Fig. 14.9 – The limit cycle.

In this way we can very quickly find the amplitude and period of these limit cycle oscillations.

Problems

1. For $F(y)$ as shown in Fig. 14.10, confirm that the amplitude of the limit cycle for large ϵ is ϵ, and the period is $2\epsilon \ln 3$.

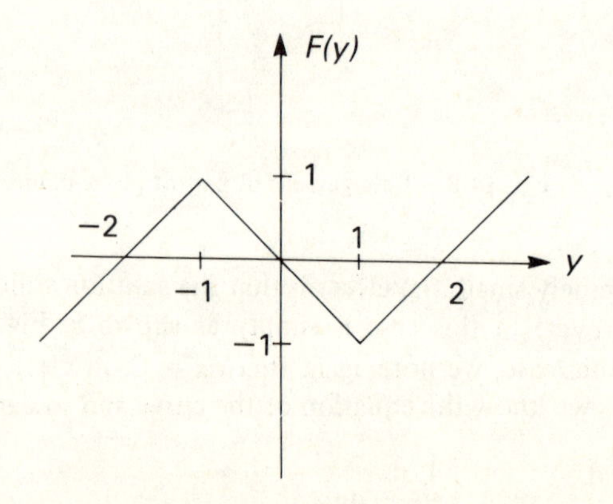

Fig. 14.10 – The nonlinearity $F(y)$.

2. For $F(y) = \frac{1}{3}y(y^2 - 3)$, confirm that the amplitude of the limit cycle for large ϵ is $\frac{2}{3}\epsilon$, and the period is $(3 - 2 \ln 2)\epsilon$. If we had used Linstedt's method for the same equation, but for the small ϵ case, the results would have been $\omega = 1 - \epsilon^2/16$. Thus $T \approx 2\pi(1 + \epsilon^2/16)$. Draw a sketch of $T \sim \epsilon$ to see how well the results for $\epsilon \to 0$ and $\epsilon \to \infty$ patch together.

3. Consider the equation

$$\ddot{x} + x + \epsilon(\dot{x}|\dot{x}| - 2\dot{x}) = 0$$

for $\epsilon \gg 1$.
Confirm that the limit cycle has amplitude ϵ and period $T \approx 2.13\epsilon$.

4. For the equation of problem 3, what happens if $\epsilon \ll -1$?

5. Consider $\ddot{x} + \epsilon F(\dot{x}) + x = 0$, where

$$F(y) = \begin{cases} y + 2, & y < -1 \\ -y & |y| \leqslant 1 \\ y - 2, & y > 1 \end{cases} .$$

(i) Explain why you expect a limit cycle.
(ii) Show that the period and amplitude of relaxation oscillations are $2\epsilon \ln 3$ and ϵ.
(iii) Use harmonic linearisation to show that the period and amplitude of the limit cycle are 1 and 2.475.

References and further reading

Bogoliubov, N. N., and Mitropolsky, Y. A., (1961) *Asymptotic Methods in the Theory of Nonlinear Oscillations,* Hindustan Press.

Braun, M., (1975) *Differential Equations and their Applications,* Springer-Verlag.

Cole, J. D., (1968) *Perturbation Methods in Applied Maths,* Blaisdell.

Dunning-Davies, J., (1982) *Mathematical Methods for Mathematicians, Physical Scientists and Engineers.* Ellis Horwood.

Jacobs, O. L. R., (1974) *Introduction to Control Theory,* Clarendon Press.

Jordan, D. W. and Smith, P., (1977) *Nonlinear Ordinary Differential Equations,* OUP.

Murdoch, D. C., (1970) *Linear Algebra,* John Wiley.

Nayfeh, A. H., (1973) *Perturbation Methods,* John Wiley.

Nayfeh, A. H. and Mook, D. T., (1979) *Nonlinear Oscillations,* John Wiley.

Oke, K., (1981), In *Case Studies in Mathematical Modelling* (D. J. G. James, J. J. McDonald, eds.), Thornes.

Strain, M., (1961) *Mathematical Methods for Technologists,* EUP.

Van Dyke, M., (1964) *Perturbation Methods in Fluid Mechanics,* Academic Press.

Index

A

amplitude response curve, 147
analytical techniques, 93
asymptotic method, 139
averaging, 158
autonomous equation, 93

B

basis, 16,19

C

castor, 175
centre, 66, 117
clockwise description, 104
closed curve in phase plane, 100, 104
continuous systems, 93
control equation, 108

D

degeneracy, 33, 42
describing function method, 166
determinant, 16, 23
dimension of a vector space, 19
diode, 137
Duffing's equation, 142, 146, 153, 166

E

eigenvalue, 22
    complex, 44
    repeated, 32
eigenvector, 22
    complex, 44
    direction, 58
equilibrium point, 63

F

focus, 65, 117
forced oscillations, 146
forced systems, 68, 79
friction, 106, 171

G

galloping of transmission lines, 172
general solution of a system, 22
graphical methods, 98

H

harmonic linearisation, 166
harmonics, 96
higher order systems, 68
homogeneous equation, 86
Hooke's law, 137
hysteresis, 148

I

inhomogeneous equations, 78
initial condition, 50
initial value problem (IVP), 50
instability, 63
integral curves, 98
interacting populations, 127
isoclines, 98

K

Kipp oscillation, 181
Krylov and Bogoliubov method, 157

L

Lanchester's tourbillion, 174
Liénard's method, 98, 100

Lindstedt's method, 142
linear approximation, 116
linear combination, 14
linear systems, 13
linearisation, 115
linearly dependent, 14
linearly independent, 14
limit cycle, 124, 155, 160, 164, 184
  stability, 163

M

matrix,
  definition, 13
  exponential, 35
  partitioned form, 79
  symmetric, 77
missile, 108
multiple timescales, 150

N

node, 64, 117
nonlinear equations, 94
nonsecular term (NST), 144
normal mode, 76

O

oscillatory solutions, 63

P

periodic motion, 100, 104
phase plane, 99, 104
Poincaré result, 117
Poincaré's method, 138

R

Rayleigh's equation, 103, 167
reduction formulae, 159
relaxation oscillation, 132, 181
relay switch, 111
resonance, 147
rotating simple pendulum, 130

S

saddle point, 63, 117
secular term, 142, 143
self-excited oscillations, 171
separatrix, 121
shimmying, 176
simple harmonic motion (SHM), 140, 150
singular point, 115
solution curve, 58
stability, 63, 117
Strouhal number, 175
switching line, 109

T

Tacoma bridge disaster, 175
Taylor series, 116
trace of a matrix, 31
trapping, 107, 109

U

undamped systems, 72

V

valve oscillator, 178
Van der Pol equation, 122, 142, 145, 149, 155, 159, 164, 168, 180
vector,
  definition, 13
  n-dimensional, 14
  space, 16
vibrating systems, 73
violin bow, 171
vortex shedding, 175